实用数据科学和
Python 机器学习(影印版)
Hands-On Data Science and Python Machine Learning

Frank Kane 著

南京　东南大学出版社

图书在版编目(CIP)数据

实用数据科学和 Python 机器学习：英文/(美)弗兰克·凯恩(Frank Kane)著. —影印本. —南京：东南大学出版社,2019.5

书名原文：Hands-On Data Science and Python Machine Learning

ISBN 978-7-5641-8320-2

Ⅰ.①实⋯ Ⅱ.①弗⋯ Ⅲ.①软件工具-程序设计-英文 Ⅳ.①TP311.561

中国版本图书馆 CIP 数据核字(2019)第 046197 号

图字：10-2018-503 号

© 2017 by PACKT Publishing Ltd.

Reprint of the English Edition, jointly published by PACKT Publishing Ltd and Southeast University Press, 2019. Authorized reprint of the original English edition, 2018 PACKT Publishing Ltd, the owner of all rights to publish and sell the same.

All rights reserved including the rights of reproduction in whole or in part in any form.

英文原版由 PACKT Publishing Ltd 出版 2018。

英文影印版由东南大学出版社出版 2019。此影印版的出版和销售得到出版权和销售权的所有者——PACKT Publishing Ltd 的许可。

版权所有，未得书面许可，本书的任何部分和全部不得以任何形式重制。

实用数据科学和 Python 机器学习（影印版）

出版发行：东南大学出版社
地　　址：南京四牌楼 2 号　　邮编：210096
出 版 人：江建中
网　　址：http://www.seupress.com
电子邮件：press@seupress.com
印　　刷：常州市武进第三印刷有限公司
开　　本：787 毫米×980 毫米　　16 开本
印　　张：26.25
字　　数：514 千字
版　　次：2019 年 5 月第 1 版
印　　次：2019 年 5 月第 1 次印刷
书　　号：ISBN 978-7-5641-8320-2
定　　价：99.00 元

本社图书若有印装质量问题，请直接与营销部联系。电话(传真)：025-83791830

Credits

Author
Frank Kane

Acquisition Editor
Ben Renow-Clarke

Content Development Editor
Khushali Bhangde

Technical Editor
Nidhisha Shetty

Copy Editor
Tom Jacob

Proofreader
Safis Editing

Indexer
Tejal Daruwale Soni

Graphics
Jason Monteiro

Production Coordinator
Arvindkumar Gupta

About the Author

My name is Frank Kane. I spent nine years at amazon.com and imdb.com, wrangling millions of customer ratings and customer transactions to produce things such as personalized recommendations for movies and products and "people who bought this also bought." I tell you, I wish we had Apache Spark back then, when I spent years trying to solve these problems there. I hold 17 issued patents in the fields of distributed computing, data mining, and machine learning. In 2012, I left to start my own successful company, Sundog Software, which focuses on virtual reality environment technology, and teaching others about big data analysis.

www.PacktPub.com

For support files and downloads related to your book, please visit www.PacktPub.com. Did you know that Packt offers eBook versions of every book published, with PDF and ePub files available? You can upgrade to the eBook version at www.PacktPub.com and as a print book customer, you are entitled to a discount on the eBook copy. Get in touch with us at service@packtpub.com for more details. At www.PacktPub.com, you can also read a collection of free technical articles, sign up for a range of free newsletters and receive exclusive discounts and offers on Packt books and eBooks.

https://www.packtpub.com/mapt

Get the most in-demand software skills with Mapt. Mapt gives you full access to all Packt books and video courses, as well as industry-leading tools to help you plan your personal development and advance your career.

Why subscribe?

- Fully searchable across every book published by Packt
- Copy and paste, print, and bookmark content
- On demand and accessible via a web browser

Customer Feedback

Thanks for purchasing this Packt book. At Packt, quality is at the heart of our editorial process. To help us improve, please leave us an honest review on this book's Amazon page at `https://www.amazon.com/dp/1787280748`.

If you'd like to join our team of regular reviewers, you can email us at `customerreviews@packtpub.com`. We award our regular reviewers with free eBooks and videos in exchange for their valuable feedback. Help us be relentless in improving our products!

Table of Contents

Preface   1

Chapter 1: Getting Started   7

  Installing Enthought Canopy   8
    Giving the installation a test run   12
      If you occasionally get problems opening your IPNYB files   15
  Using and understanding IPython (Jupyter) Notebooks   15
  Python basics - Part 1   20
  Understanding Python code   22
  Importing modules   25
    Data structures   26
    Experimenting with lists   26
      Pre colon   27
      Post colon   27
      Negative syntax   28
      Adding list to list   28
      The append function   28
      Complex data structures   29
      Dereferencing a single element   29
      The sort function   30
      Reverse sort   30
    Tuples   30
      Dereferencing an element   31
      List of tuples   31
      Dictionaries   33
    Iterating through entries   34
  Python basics - Part 2   35
    Functions in Python   35
      Lambda functions - functional programming   36
      Understanding boolean expressions   37
        The if statement   37
        The if-else loop   38
    Looping   38
      The while loop   39
    Exploring activity   40
  Running Python scripts   41
    More options than just the IPython/Jupyter Notebook   42
    Running Python scripts in command prompt   43

Using the Canopy IDE	44
Summary	47
Chapter 2: Statistics and Probability Refresher, and Python Practice	49
Types of data	50
Numerical data	51
Discrete data	51
Continuous data	51
Categorical data	52
Ordinal data	53
Mean, median, and mode	54
Mean	55
Median	55
The factor of outliers	56
Mode	57
Using mean, median, and mode in Python	58
Calculating mean using the NumPy package	58
Visualizing data using matplotlib	59
Calculating median using the NumPy package	60
Analyzing the effect of outliers	61
Calculating mode using the SciPy package	62
Some exercises	64
Standard deviation and variance	65
Variance	66
Measuring variance	67
Standard deviation	68
Identifying outliers with standard deviation	69
Population variance versus sample variance	70
The Mathematical explanation	70
Analyzing standard deviation and variance on a histogram	72
Using Python to compute standard deviation and variance	73
Try it yourself	74
Probability density function and probability mass function	74
The probability density function and probability mass functions	74
Probability density functions	75
Probability mass functions	76
Types of data distributions	77
Uniform distribution	77
Normal or Gaussian distribution	78
The exponential probability distribution or Power law	80
Binomial probability mass function	82
Poisson probability mass function	83

Percentiles and moments	84
Percentiles	84
Quartiles	86
Computing percentiles in Python	87
Moments	90
Computing moments in Python	93
Summary	95
Chapter 3: Matplotlib and Advanced Probability Concepts	97
A crash course in Matplotlib	98
Generating multiple plots on one graph	99
Saving graphs as images	100
Adjusting the axes	101
Adding a grid	102
Changing line types and colors	103
Labeling axes and adding a legend	107
A fun example	108
Generating pie charts	110
Generating bar charts	111
Generating scatter plots	112
Generating histograms	113
Generating box-and-whisker plots	113
Try it yourself	115
Covariance and correlation	116
Defining the concepts	116
Measuring covariance	117
Correlation	118
Computing covariance and correlation in Python	118
Computing correlation – The hard way	118
Computing correlation – The NumPy way	122
Correlation activity	124
Conditional probability	124
Conditional probability exercises in Python	125
Conditional probability assignment	129
My assignment solution	130
Bayes' theorem	132
Summary	134
Chapter 4: Predictive Models	135
Linear regression	135
The ordinary least squares technique	137

The gradient descent technique	138
The co-efficient of determination or r-squared	139
Computing r-squared	139
Interpreting r-squared	139
Computing linear regression and r-squared using Python	140
Activity for linear regression	143
Polynomial regression	**144**
Implementing polynomial regression using NumPy	145
Computing the r-squared error	148
Activity for polynomial regression	148
Multivariate regression and predicting car prices	**149**
Multivariate regression using Python	151
Activity for multivariate regression	154
Multi-level models	**155**
Summary	**157**
Chapter 5: Machine Learning with Python	**159**
Machine learning and train/test	**159**
Unsupervised learning	160
Supervised learning	161
Evaluating supervised learning	162
K-fold cross validation	164
Using train/test to prevent overfitting of a polynomial regression	**164**
Activity	170
Bayesian methods - Concepts	**170**
Implementing a spam classifier with Naïve Bayes	**172**
Activity	176
K-Means clustering	**177**
Limitations to k-means clustering	179
Clustering people based on income and age	**181**
Activity	184
Measuring entropy	**184**
Decision trees - Concepts	**186**
Decision tree example	188
Walking through a decision tree	190
Random forests technique	190
Decision trees - Predicting hiring decisions using Python	**191**
Ensemble learning – Using a random forest	197
Activity	198
Ensemble learning	**198**
Support vector machine overview	**201**

Using SVM to cluster people by using scikit-learn	203
Activity	207
Summary	208

Chapter 6: Recommender Systems — 209

What are recommender systems?	210
User-based collaborative filtering	212
Limitations of user-based collaborative filtering	214
Item-based collaborative filtering	215
Understanding item-based collaborative filtering	215
How item-based collaborative filtering works?	216
Collaborative filtering using Python	220
Finding movie similarities	220
Understanding the code	222
The corrwith function	225
Improving the results of movie similarities	228
Making movie recommendations to people	233
Understanding movie recommendations with an example	238
Using the groupby command to combine rows	240
Removing entries with the drop command	241
Improving the recommendation results	242
Summary	244

Chapter 7: More Data Mining and Machine Learning Techniques — 245

K-nearest neighbors - concepts	246
Using KNN to predict a rating for a movie	248
Activity	255
Dimensionality reduction and principal component analysis	256
Dimensionality reduction	256
Principal component analysis	257
A PCA example with the Iris dataset	259
Activity	264
Data warehousing overview	264
ETL versus ELT	266
Reinforcement learning	268
Q-learning	269
The exploration problem	270
The simple approach	270
The better way	271
Fancy words	271
Markov decision process	271
Dynamic programming	272

| Summary | 274 |

Chapter 8: Dealing with Real-World Data — 275

Bias/variance trade-off	276
K-fold cross-validation to avoid overfitting	279
Example of k-fold cross-validation using scikit-learn	280
Data cleaning and normalisation	284
Cleaning web log data	287
Applying a regular expression on the web log	288
Modification one - filtering the request field	291
Modification two - filtering post requests	293
Modification three - checking the user agents	295
Filtering the activity of spiders/robots	297
Modification four - applying website-specific filters	299
Activity for web log data	301
Normalizing numerical data	301
Detecting outliers	303
Dealing with outliers	304
Activity for outliers	307
Summary	307

Chapter 9: Apache Spark - Machine Learning on Big Data — 309

Installing Spark	310
Installing Spark on Windows	310
Installing Spark on other operating systems	311
Installing the Java Development Kit	312
Installing Spark	319
Spark introduction	333
It's scalable	334
It's fast	335
It's young	336
It's not difficult	336
Components of Spark	336
Python versus Scala for Spark	337
Spark and Resilient Distributed Datasets (RDD)	338
The SparkContext object	339
Creating RDDs	340
Creating an RDD using a Python list	340
Loading an RDD from a text file	340
More ways to create RDDs	341
RDD operations	341

Transformations	342
Using map()	343
Actions	343
Introducing MLlib	**344**
Some MLlib Capabilities	345
Special MLlib data types	345
The vector data type	346
LabeledPoint data type	346
Rating data type	346
Decision Trees in Spark with MLlib	**347**
Exploring decision trees code	348
Creating the SparkContext	349
Importing and cleaning our data	351
Creating a test candidate and building our decision tree	356
Running the script	357
K-Means Clustering in Spark	**359**
Within set sum of squared errors (WSSSE)	363
Running the code	364
TF-IDF	**365**
TF-IDF in practice	366
Using TF-IDF	367
Searching wikipedia with Spark MLlib	**367**
Import statements	369
Creating the initial RDD	369
Creating and transforming a HashingTF object	370
Computing the TF-IDF score	371
Using the Wikipedia search engine algorithm	371
Running the algorithm	372
Using the Spark 2.0 DataFrame API for MLlib	**373**
How Spark 2.0 MLlib works	373
Implementing linear regression	374
Summary	**378**
Chapter 10: Testing and Experimental Design	**379**
A/B testing concepts	**379**
A/B tests	379
Measuring conversion for A/B testing	382
How to attribute conversions	383
Variance is your enemy	383
T-test and p-value	**384**
The t-statistic or t-test	385
The p-value	385

Measuring t-statistics and p-values using Python 387
Running A/B test on some experimental data 387
When there's no real difference between the two groups 388
Does the sample size make a difference? 389
Sample size increased to six-digits 389
Sample size increased seven-digits 390
A/A testing 390
Determining how long to run an experiment for 391
A/B test gotchas 392
Novelty effects 394
Seasonal effects 394
Selection bias 395
Auditing selection bias issues 396
Data pollution 396
Attribution errors 397
Summary 397

Index 399

Preface

Being a data scientist in the tech industry is one of the most rewarding careers on the planet today. I went and studied actual job descriptions for data scientist roles at tech companies and I distilled those requirements down into the topics that you'll see in this course.

Hands-On Data Science and Python Machine Learning is really comprehensive. We'll start with a crash course on Python and do a review of some basic statistics and probability, but then we're going to dive right into over 60 topics in data mining and machine learning. That includes things such as Bayes' theorem, clustering, decision trees, regression analysis, experimental design; we'll look at them all. Some of these topics are really fun.

We're going to develop an actual movie recommendation system using actual user movie rating data. We're going to create a search engine that actually works for Wikipedia data. We're going to build a spam classifier that can correctly classify spam and nonspam emails in your email account, and we also have a whole section on scaling this work up to a cluster that runs on big data using Apache Spark.

If you're a software developer or programmer looking to transition into a career in data science, this course will teach you the hottest skills without all the mathematical notation and pretense that comes along with these topics. We're just going to explain these concepts and show you some Python code that actually works that you can dive in and mess around with to make those concepts sink home, and if you're working as a data analyst in the finance industry, this course can also teach you to make the transition into the tech industry. All you need is some prior experience in programming or scripting and you should be good to go.

The general format of this book is I'll start with each concept, explaining it in a bunch of sections and graphical examples. I will introduce you to some of the notations and fancy terminologies that data scientists like to use so you can talk the same language, but the concepts themselves are generally pretty simple. After that, I'll throw you into some actual Python code that actually works that we can run and mess around with, and that will show you how to actually apply these ideas to actual data. These are going to be presented as IPython Notebook files, and that's a format where I can intermix code and notes surrounding the code that explain what's going on in the concepts. You can take these notebook files with you after going through this book and use that as a handy-quick reference later on in your career, and at the end of each concept, I'll encourage you to actually dive into that Python code, make some modifications, mess around with it, and just gain more familiarity by getting hands-on and actually making some modifications, and seeing the effects they have.

Who this book is for

If you are a budding data scientist or a data analyst who wants to analyze and gain actionable insights from data using Python, this book is for you. Programmers with some experience in Python who want to enter the lucrative world of Data Science will also find this book to be very useful.

Conventions

In this book, you will find a number of text styles that distinguish between different kinds of information. Here are some examples of these styles and an explanation of their meaning.

Code words in text, database table names, folder names, filenames, file extensions, pathnames, dummy URLs, user input, and Twitter handles are shown as follows: "We can measure that using the `r2_score()` function from `sklearn.metrics`."

A block of code is set as follows:

```
import numpy as np
import pandas as pd
from sklearn import tree

input_file = "c:/spark/DataScience/PastHires.csv"
df = pd.read_csv(input_file, header = 0)
```

When we wish to draw your attention to a particular part of a code block, the relevant lines or items are set in bold:

```
import numpy as np
import pandas as pd
from sklearn import tree

input_file = "c:/spark/DataScience/PastHires.csv"
df = pd.read_csv(input_file, header = 0)
```

Any command-line input or output is written as follows:

```
spark-submit SparkKMeans.py
```

Preface

New terms and important words are shown in bold. Words that you see on the screen, for example, in menus or dialog boxes, appear in the text like this: "On Windows 10, you'll need to open up the **Start** menu and go to **Windows System** | **Control Panel** to open up **Control Panel**."

 Warnings or important notes appear like this.

 Tips and tricks appear like this.

Reader feedback

Feedback from our readers is always welcome. Let us know what you think about this book-what you liked or disliked. Reader feedback is important for us as it helps us develop titles that you will really get the most out of.

To send us general feedback, simply email `feedback@packtpub.com`, and mention the book's title in the subject of your message.

If there is a topic that you have expertise in and you are interested in either writing or contributing to a book, see our author guide at `www.packtpub.com/authors`.

Customer support

Now that you are the proud owner of a Packt book, we have a number of things to help you to get the most from your purchase.

Downloading the example code

You can download the example code files for this book from your account at `http://www.packtpub.com`. If you purchased this book elsewhere, you can visit `http://www.packtpub.com/support` and register to have the files emailed directly to you.

You can download the code files by following these steps:

1. Log in or register to our website using your email address and password.
2. Hover the mouse pointer on the **SUPPORT** tab at the top.
3. Click on **Code Downloads & Errata**.
4. Enter the name of the book in the **Search** box.
5. Select the book for which you're looking to download the code files.
6. Choose from the drop-down menu where you purchased this book from.
7. Click on **Code Download**.

You can also download the code files by clicking on the Code Files button on the book's webpage at the Packt Publishing website. This page can be accessed by entering the book's name in the Search box. Please note that you need to be logged in to your Packt account.

Once the file is downloaded, please make sure that you unzip or extract the folder using the latest version of:

- WinRAR / 7-Zip for Windows
- Zipeg / iZip / UnRarX for Mac
- 7-Zip / PeaZip for Linux

The code bundle for the book is also hosted on GitHub at https://github.com/PacktPublishing/Hands-On-Data-Science-and-Python-Machine-Learning. We also have other code bundles from our rich catalog of books and videos available at https://github.com/PacktPublishing/. Check them out!

Downloading the color images of this book

We also provide you with a PDF file that has color images of the screenshots/diagrams used in this book. The color images will help you better understand the changes in the output. You can download this file from https://www.packtpub.com/sites/default/files/downloads/HandsOnDataScienceandPythonMachineLearning_ColorImages.pdf.

Errata

Although we have taken every care to ensure the accuracy of our content, mistakes do happen. If you find a mistake in one of our books-maybe a mistake in the text or the code- we would be grateful if you could report this to us. By doing so, you can save other readers from frustration and help us improve subsequent versions of this book. If you find any errata, please report them by visiting http://www.packtpub.com/submit-errata, selecting your book, clicking on the Errata Submission Form link, and entering the details of your errata. Once your errata are verified, your submission will be accepted and the errata will be uploaded to our website or added to any list of existing errata under the Errata section of that title.

To view the previously submitted errata, go to https://www.packtpub.com/books/content/support and enter the name of the book in the search field. The required information will appear under the Errata section.

Piracy

Piracy of copyrighted material on the internet is an ongoing problem across all media. At Packt, we take the protection of our copyright and licenses very seriously. If you come across any illegal copies of our works in any form on the internet, please provide us with the location address or website name immediately so that we can pursue a remedy.

Please contact us at copyright@packtpub.com with a link to the suspected pirated material.

We appreciate your help in protecting our authors and our ability to bring you valuable content.

Questions

If you have a problem with any aspect of this book, you can contact us at questions@packtpub.com, and we will do our best to address the problem.

1
Getting Started

Since there's going to be code associated with this book and sample data that you need to get as well, let me first show you where to get that and then we'll be good to go. We need to get some setup out of the way first. First things first, let's get the code and the data that you need for this book so you can play along and actually have some code to mess around with. The easiest way to do that is by going right to this - *Getting Started*.

In this chapter, we will first install and get ready in a working Python environment:

- Installing Enthought Canopy
- Installing Python libraries
- How to work with the IPython/Jupyter Notebook
- How to use, read and run the code files for this book
- Then we'll dive into a crash course into understanding Python code:
- Python basics - part 1
- Understanding Python code
- Importing modules
- Experimenting with lists
- Tuples
- Python basics - part 2
- Running Python scripts

You'll have everything you need for an amazing journey into data science with Python, once we've set up your environment and familiarized you with Python in this chapter.

Getting Started

Installing Enthought Canopy

Let's dive right in and get what you need installed to actually develop Python code with data science on your desktop. I'm going to walk you through installing a package called Enthought Canopy which has both the development environment and all the Python packages you need pre-installed. It makes life really easy, but if you already know Python you might have an existing Python environment already on your PC, and if you want to keep using it, maybe you can.

The most important thing is that your Python environment has Python 3.5 or newer, that it supports Jupyter Notebooks (because that's what we're going to use in this course), and that you have the key packages you need for this book installed on your environment. I'll explain exactly how to achieve a full installation in a few simple steps - it's going to be very easy.

Let's first overview those key packages, most of which Canopy will be installing for us automatically for us. Canopy will install Python 3.5 for us, and some further packages we need including: `scikit_learn`, `xlrd`, and `statsmodels`. We'll need to manually use the `pip` command, to install a package called `pydot2plus`. And that will be it - it's very easy with Canopy!

Once the following installation steps are complete, we'll have everything we need to actually get up and running, and so we'll open up a little sample file and do some data science for real. Now let's get you set up with everything you need to get started as quickly as possible:

> 1. The first thing you will need is a development environment, called an IDE, for Python code. What we're going to use for this book is Enthought Canopy. It's a scientific computing environment, and it's going to work well with this book:

2. To get Canopy installed, just go to www.enthought.com and click on **DOWNLOADS: Canopy**:

Getting Started

3. Enthought Canopy is free, for the Canopy Express edition - which is what you want for this book. You must then select your operating system and architecture. For me, that's Windows 64-bit, but you'll want to click on corresponding Download button for your operating system and with the Python 3.5 option:

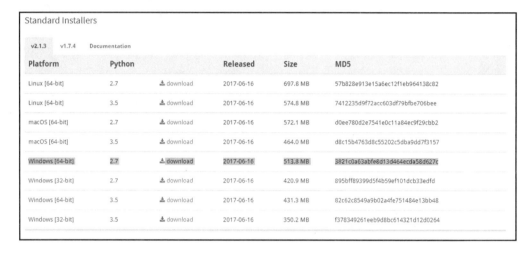

4. We don't have to give them any personal information at this step. There's a pretty standard Windows installer, so just let that download:

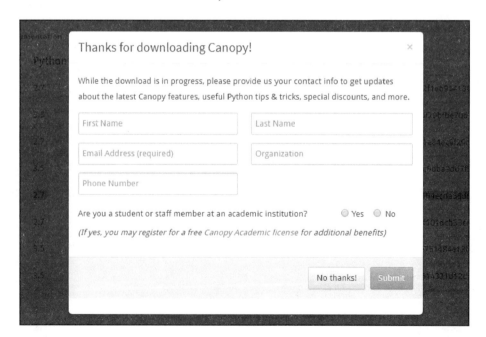

5. After that's downloaded we go ahead and open up the Canopy installer, and run it! You might want to read the license before you agree to it, that's up to you, and then just wait for the installation to complete.
6. Once you hit the **Finish** button at the end of the install process, allow it to launch Canopy automatically. You'll see that Canopy then sets up the Python environment by itself, which is great, but this will take a minute or two.
7. Once the installer is done setting up your Python environment, you should get a screen that looks like the one below. It says welcome to Canopy and a bunch of big friendly buttons:

Getting Started

8. The beautiful thing is that pretty much everything you need for this book comes pre-installed with Enthought Canopy, that's why I recommend using it!
9. There is just one last thing we need to set up, so go ahead and click the Editor button there on the Canopy Welcome screen. You'll then see the Editor screen come up, and if you click down in the window at the bottom, I want you to just type in:

   ```
   !pip install pydotplus
   ```

10. Here's how that's going to look on your screen as you type the above line in at the bottom of the Canopy Editor window; don't forget to press the Return button of course:

    ```
    %quickref  -> Quick reference.
    help       -> Python's own help system.
    object?    -> Details about 'object', use 'objec

    In [1]: !pip install pydotplus
    ```

11. One you hit the Return button, this will install that one extra module that we need for later on in the book, when we get to talking about decision trees, and rendering decision trees.
12. Once it has finished installing **pydotplus**, it should come back and say it's successfully installed and, voila, you have everything you need now to get started! The installation is done, at this point - but let's just take a few more steps to confirm our installation is running nicely.

Giving the installation a test run

1. Let's now give your installation a test run. The first thing to do is actually to entirely close the Canopy window! This is because we're not actually going to be editing and using our code within this Canopy editor. Instead we're going to be using something called an IPython Notebook, which is also now known as the Jupyter Notebook.

2. Let me show you how that works. If you now open a window in your operating system to view the accompanying book files that you downloaded, as described in the Preface of this book. It should look something like this, with the set of .ipynb code files you downloaded for this book:

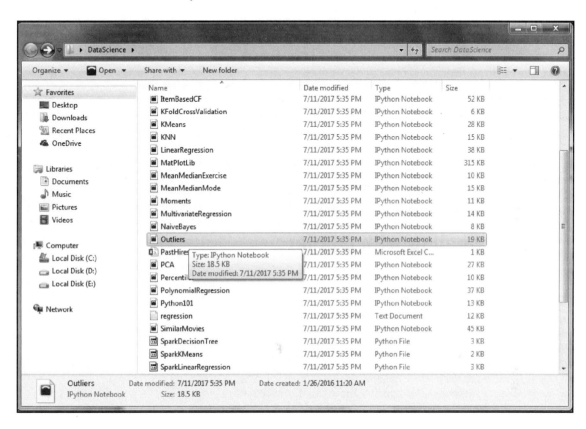

Getting Started

Now go down to the Outliers file in the list, that's the `Outliers.ipynb` file, double-click it, and what should happen is it's going to start up Canopy first and then it's going to kick off your web browser! This is because IPython/Jupyter Notebooks actually live within your web browser. There can be a small pause at first, and it can be a little bit confusing first time, but you'll soon get used to the idea.

You should soon see Canopy come up and for me my default web browser Chrome comes up. You should see the following Jupyter Notebook page, since we double-clicked on the `Outliers.ipynb` file:

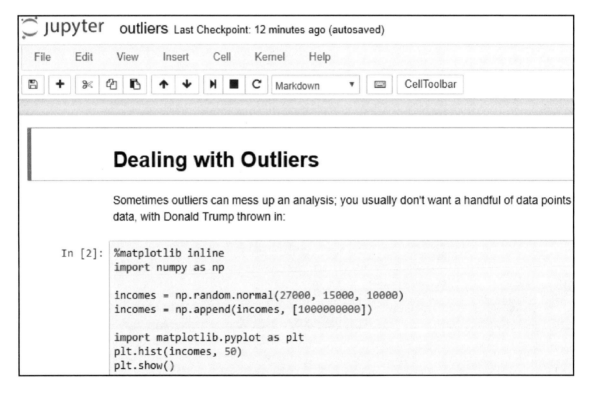

If you see this screen, it means that everything's working great in your installation and you're all set for the journey across rest of this book!

If you occasionally get problems opening your IPNYB files

Just occasionally, I've noticed that things can go a little bit wrong when you double-click on a `.ipynb` file. Don't panic! Just sometimes, Canopy can get a little bit flaky, and you might see a screen that is looking for some password or token, or you might occasionally see a screen that says it can't connect at all.

Don't panic if either of those things happen to you, they are just random quirks, sometimes things just don't start up in the right order or they don't start up in time on your PC and it's okay.

All you have to do is go back and try to open that file a second time. Sometimes it takes two or three tries to actually get it loaded up properly, but if you do it a couple of times it should pop up eventually, and a Jupyter Notebook screen like the one we saw previously about **Dealing with Outliers**, is what you should see.

Using and understanding IPython (Jupyter) Notebooks

Congratulations on your installation! Let's now explore using Jupyter Notebooks, which is also known as IPython Notebook. These days, the more modern name is the Jupyter Notebook, but a lot of people still call it an IPython Notebook, and I consider the names interchangeable for working developers as a result. I do also find the name IPython Notebooks helps me remember the notebook file name suffix which is `.ipynb` as you'll get to know very well in this book!

Okay so now let's take it right from the top again - with our first exploration of the IPython/Jupyter Notebook. If you haven't yet done so, please navigate to the `DataScience` folder where we have downloaded all the materials for this book. For me, that's `E:DataScience`, and if you didn't do so during the preceding installation section, please now double-click and open up the `Outliers.ipynb` file.

Getting Started

Now what's going to happen when we double-click on this IPython .ipynb file is that first of all it's going to spark up Canopy, if it's not sparked up already, and then it's going to launch a web browser. This is how the full Outliers notebook webpage looks within my browser:

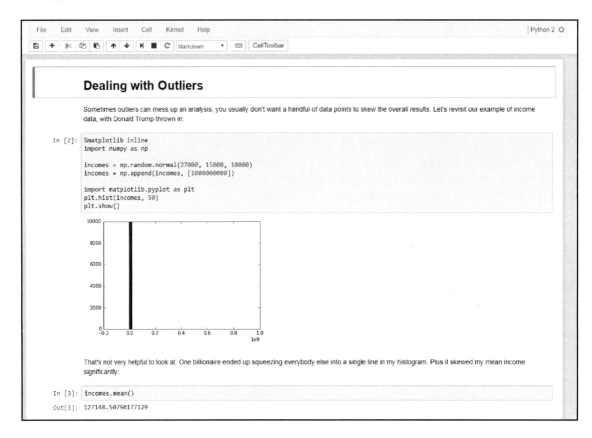

As you can see here, notebooks are structured in such a way that I can intersperse my little notes and commentary about what you're seeing here within the actual code itself, and you can actually run this code within your web browser! So, it's a very handy format for me to give you sort of a little reference that you can use later on in life to go and remind yourself how these algorithms work that we're going to talk about, and actually experiment with them and play with them yourself.

The way that the IPython/Jupyter Notebook files work is that they actually run from within your browser, like a webpage, but they're backed by the Python engine that you installed. So you should be seeing a screen similar to the one shown in the previous screenshot.

You'll notice as you scroll down the notebook in your browser, there are code blocks. They're easy to spot because they contain our actual code. Please find the code box for this code in the Outliers notebook, quite near the top:

```
%matplotlib inline
import numpy as np

incomes = np.random.normal(27000, 15000, 10000)
incomes = np.append(incomes, [1000000000])

import matplotlib.pyplot as plt
plt.hist(incomes, 50)
plt.show()
```

Let's take a quick look at this code while we're here. We are setting up a little income distribution in this code. We're simulating the distribution of income in a population of people, and to illustrate the effect that an outlier can have on that distribution, we're simulating Donald Trump entering the mix and messing up the mean value of the income distribution. By the way, I'm not making a political statement, this was all done before Trump became a politician. So you know, full disclosure there.

We can select any code block in the notebook by clicking on it. So if you now click in the code block that contains the code we just looked at above, we can then hit the run button at the top to run it. Here's the area at the top of the screen where you'll find the Run button:

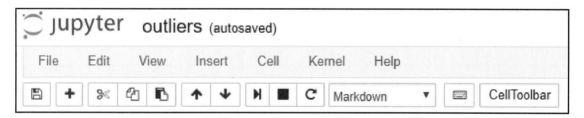

Getting Started

Hitting the Run button with the code block selected, will cause this graph to be regenerated:

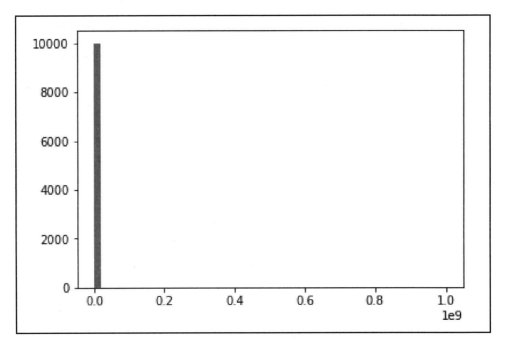

Similarly, we can click on the next code block a little further down, you'll spot the one which has the following single line of code :

```
incomes.mean()
```

If you select the code block containing this line, and hit the Run button to run the code, you'll see the output below it, which ends up being a very large value because of the effect of that outlier, something like this:

```
127148.50796177129
```

Let's keep going and have some fun. In the next code block down, you'll see the following
code, which tries to detect outliers like Donald Trump and remove them from the dataset:

```
def reject_outliers(data):
    u = np.median(data)
    s = np.std(data)
    filtered = [e for e in data if (u - 2 * s < e < u + 2 * s)]
    return filtered

filtered = reject_outliers(incomes)
plt.hist(filtered, 50)
plt.show()
```

So select the corresponding code block in the notebook, and press the run button again.
When you do that, you'll see this graph instead:

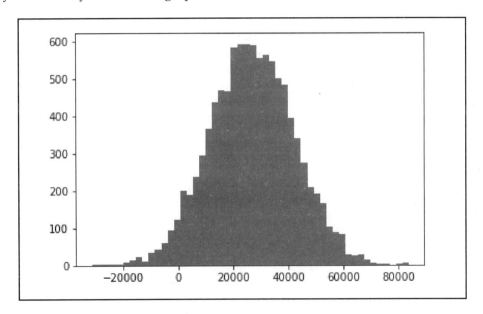

Now we see a much better histogram that represents the more typical American - now that
we've taken out our outlier that was messing things up.

So, at this point, you have everything you need to get started in this course. We have all the
data you need, all the scripts, and the development environment for Python and Python
notebooks. So, let's rock and roll. Up next we're going to do a little crash course on Python
itself, and even if you're familiar with Python, it might be a good little refresher so you
might want to watch it regardless. Let's dive in and learn Python.

Python basics - Part 1

If you already know Python, you can probably skip the next two sections. However, if you need a refresher, or if you haven't done Python before, you'll want to go through these. There are a few quirky things about the Python scripting language that you need to know, so let's dive in and just jump into the pool and learn some Python by writing some actual code.

Like I said before, in the requirements for this book, you should have some sort of programming background to be successful in this book. You've coded in some sort of language, even if it's a scripting language, JavaScript, I don't care whether it is C++, Java, or something, but if you're new to Python, I'm going to give you a little bit of a crash course here. I'm just going to dive right in and go right into some examples in this section.

There are a few quirks about Python that are a little bit different than other languages you might have seen; so I just want to walk through what's different about Python from other scripting languages you may have worked with, and the best way to do that is by looking at some real examples. Let's dive right in and look at some Python code:

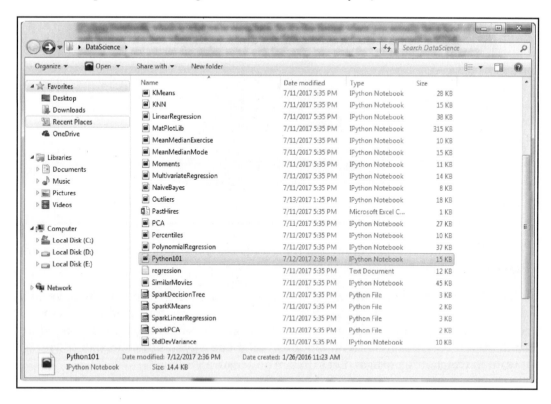

Chapter 1

If you open up the DataScience folder for this class, which you downloaded earlier in the earlier section, you should find a Python101.ipynb file; go ahead and double-click on that. It should open right up in Canopy if you have everything installed properly, and it should look a little bit something like the following screenshot:

Python Basics

Whitespace Is Important

```
In [1]: listOfNumbers = [1, 2, 3, 4, 5, 6]

        for number in listOfNumbers:
            print (number),
            if (number % 2 == 0):
                print ("is even")
            else:
                print ("is odd")

        print ("All done.")

1
is odd
2
is even
3
is odd
4
is even
5
is odd
6
is even
All done.
```

New versions of Canopy will open the code in your web browser, not the Canopy editor! This is okay!

One cool thing about Python is that there are several ways to run code with Python. You can run it as a script, like you would with a normal programming language. You can also write in this thing called the *IPython Notebook*, which is what we're using here. So it's this format where you actually have a web browser-like view where you can actually write little notations and notes to yourself in HTML markup stuff, and you can also embed actual code that really runs using the Python interpreter.

Understanding Python code

The first example that I want to give you of some Python code is right here. The following block of code represents some real Python code that we can actually run right within this view of the entire notebook page, but let's zoom in now and look at that code:

Python Basics

Whitespace Is Important

```
In [1]: listOfNumbers = [1, 2, 3, 4, 5, 6]

        for number in listOfNumbers:
            print (number),
            if (number % 2 == 0):
                print ("is even")
            else:
                print ("is odd")

        print ("All done.")
```

```
1
is odd
2
is even
3
is odd
4
is even
5
is odd
6
is even
All done.
```

Chapter 1

Let's take a look at what's going on. We have a list of numbers and a list in Python, kind of like an array in other languages. It is designated by these square brackets:

Whitespace Is Important

```
In [4]: listOfNumbers = [1, 2, 3, 4, 5, 6]
```

We have this data structure of a list that contains the numbers 1 through 6, and then to iterate through every number in that list, we'll say `for number in listOfNumbers:`, that's the Python syntax for iterating through a list of stuff and a colon.

 Tabs and whitespaces have real meaning in Python, so you can't just format things the way you want to. You have to pay attention to them.

The point that I want to make is that in other languages, it's pretty typical to have a bracket or a brace of some sort there to denote that I'm inside a `for` loop, an `if` block, or some sort of block of code, but in Python, that's all designated with whitespaces. Tab is actually important in telling Python what's in which block of code:

```
for number in listOfNumbers:
    print number,
    if (number % 2 == 0):
        print ("is even")
    else:
        print ("is odd")
print ("Hooray! We're all done.")
```

[23]

You'll notice that within this `for` block, we have a tab of one within that entire block, and for every `number in listOfNumbers` we will execute all of this code that's tabbed in by one *Tab* stop. We'll print the number, and the comma just means that we're not going to do a new line afterwards. We'll print something else right after it, and `if (number % 2 = 0)`, we'll say it's `even`. Otherwise, we'll say it's `odd`, and when we're done, we'll print out `All done`:

```
1 is odd
2 is even
3 is odd
4 is even
5 is odd
6 is even
All done.
```

You can see the output right below the code. I ran the output before as I had actually saved it within my notebook, but if you want to actually run it yourself, you can just click within that block and click on the Play button, and we'll actually execute it and do it again. Just to convince yourself that it's really doing something, let's change the `print` statement to say something else, say, `Hooray! We're all done. Let's party!` If I run this now, you can see, sure enough, my message there has changed:

Python Basics

Whitespace Is Important

```
In [9]: listOfNumbers = [1, 2, 3, 4, 5, 6]

        for number in listOfNumbers:
            print (number),
            if (number % 2 == 0):
                print ("is even")
            else:
                print ("is odd")

        print ("Hooray! We're all done. Let's party!")
```

```
1
is odd
2
is even
3
is odd
4
is even
5
is odd
6
is even
Hooray! We're all done. Let's party!
```

So again, the point I want to make is that whitespace is important. You will designate blocks of code that run together, you know, such as a `for` loop or `if then` statements, using indentation or tabs, so remember that. Also, pay attention to your colons too. You'll notice that a lot of these clauses begin with a colon.

Importing modules

Python itself, like any language, is fairly limited in what it can do. The real power of using Python for machine learning and data mining and data science is the power of all the external libraries that are available for it for that purpose. One of those libraries is called `NumPy`, or numeric Python, and, for example, here we can `import` the `Numpy` package, which is included with Canopy as `np`.

This means that I'll refer to the `NumPy` package as `np`, and I could call that anything I want. I could call it `Fred` or `Tim`, but it's best to stick with something that actually makes sense; now that I'm calling that `NumPy` package `np`, I can refer to it using `np`:

```
import numpy as np
```

In this example, I'll call the `random` function that's provided as part of the `NumPy` package and call its normal function to actually generate a normal distribution of random numbers using these parameters and print them out. Since it is random, I should get different results every time:

```
import numpy as np
A = np.random.normal(25.0, 5.0, 10)
print (A)
```

The output should look like this:

```
[ 23.50119237  28.3470395   27.68512972  27.43957344  22.66626262
  25.98055199  27.87395644  25.99525487  20.36318406  22.77226693]
```

Sure enough, I get different results. That's pretty cool.

Data structures

Let's move on to data structures. If you need to pause and let things sink in a little bit, or you want to play around with these a little bit more, feel free to do so. The best way to learn this stuff is to dive in and actually experiment, so I definitely encourage doing that, and that's why I'm giving you working IPython/Jupyter Notebooks, so you can actually go in, mess with the code, do different stuff with it.

For example, here we have a distribution around 25.0, but let's make it around 55.0:

```
import numpy as np
A = np.random.normal(55.0, 5.0, 10)
print (A)
```

Hey, all my numbers changed, they're closer to 55 now, how about that?

```
[ 48.79441876  63.77818473  61.24157056  47.38182128  52.5623337
  55.80574543  55.16594437  53.59688042  50.57639509  60.44058303]
```

Alright, let's talk about data structures a little bit here. As we saw in our first example, you can have a list, and the syntax looks like this.

Experimenting with lists

```
x = [1, 2, 3, 4, 5, 6]
print (len(x))
```

You can say, call a list x, for example, and assign it to the numbers 1 through 6, and these square brackets indicate that we are using a Python list, and those are immutable objects that I can actually add things to and rearrange as much as I want to. There's a built-in function for determining the length of the list called `len`, and if I type in `len(x)`, that will give me back the number 6 because there are 6 numbers in my list.

Just to make sure, and again to drive home the point that this is actually running real code here, let's add another number in there, such as 4545. If you run this, you'll get 7 because now there are 7 numbers in that list:

```
x = [1, 2, 3, 4, 5, 6, 4545]
print (len(x))
```

The output of the previous code example is as follows:

```
7
```

Go back to the original example there. Now you can also slice lists. If you want to take a subset of a list, there's a very simple syntax for doing so:

```
x[3:]
```

The output of the above code example is as follows:

```
[1, 2, 3]
```

Pre colon

If, for example, you want to take the first three elements of a list, everything before element number 3, we can say :3 to get the first three elements, 1, 2, and 3, and if you think about what's going on there, as far as indices go, like in most languages, we start counting from 0. So element 0 is 1, element 1 is 2, and element 2 is 3. Since we're saying we want everything before element 3, that's what we're getting.

> So, you know, never forget that in most languages, you start counting at 0 and not 1.

Now this can confuse matters, but in this case, it does make intuitive sense. You can think of that colon as meaning I want everything, I want the first three elements, and I could change that to four just again to make the point that we're actually doing something real here:

```
x[:4]
```

The output of the above code example is as follows:

```
[1, 2, 3, 4]
```

Post colon

Now if I put the colon on the other side of the 3, that says I want everything after 3, so 3 and after. If I say x[3:], that's giving me the third element, 0, 1, 2, 3, and everything after it. So that's going to return 4, 5, and 6 in that example, OK?

```
x[3:]
```

The output is as follows:

```
[4, 5, 6]
```

You might want to keep this IPython/Jupyter Notebook file around. It's a good reference, because sometimes it can get confusing as to whether the slicing operator includes that element or if it's up to or including it or not. So the best way is to just play around with it here and remind yourself.

Negative syntax

One more thing you can do is have this negative syntax:

```
x[-2:]
```

The output is as follows:

```
[5, 6]
```

By saying `x[-2:]`, this means that I want the last two elements in the list. This means that go backwards two from the end, and that will give me 5 and 6, because those are the last two things on my list.

Adding list to list

You can also change lists around. Let's say I want to add a list to the list. I can use the `extend` function for that, as shown in the following code block:

```
x.extend([7,8])
x
```

The output of the above code is as follows:

```
[1, 2, 3, 4, 5, 6, 7, 8]
```

I have my list of 1, 2, 3, 4, 5, 6. If I want to extend it, I can say I have a new list here, [7, 8], and that bracket indicates this is a new list of itself. This could be a list implicit, you know, that's inline there, it could be referred to by another variable. You can see that once I do that, the new list I get actually has that list of 7, 8 appended on to the end of it. So I have a new list by extending that list with another list.

The append function

If you want to just add one more thing to that list, you can use the `append` function. So I just want to stick the number 9 at the end, there we go:

```
x.append(9)
x
```

The output of the above code is as follows:

```
[1, 2, 3, 4, 5, 6, 7, 8, 9]
```

Complex data structures

You can also have complex data structures with lists. So you don't have to just put numbers in it; you can actually put strings in it. You can put numbers in it. You can put other lists in it. It doesn't matter. Python is a weakly-typed language, so you can pretty much put whatever kind of data you want, wherever you want, and it will generally be an OK thing to do:

```
y = [10, 11, 12]
listOfLists = [x, y]
listOfLists
```

In the preceding example, I have a second list that contains 10, 11, 12, that I'm calling y. I'll create a new list that contains two lists. How's that for mind blowing? Our `listofLists` list will contain the x list and the y list, and that's a perfectly valid thing to do. You can see here that we have a bracket indicating the `listofLists` list, and within that, we have another set of brackets indicating each individual list that is in that list:

```
[[ 1, 2, 3, 4, 5, 6, 7, 8, 9 ], [10, 11, 12]]
```

So, sometimes things like these will come in handy.

Dereferencing a single element

If you want to dereference a single element of the list you can just use the bracket like that:

```
y[1]
```

The output of the above code is as follows:

```
11
```

Getting Started

So `y[1]` will return element 1. Remember that `y` had 10, 11, 12 in it - observe the previous example, and we start counting from 0, so element 1 will actually be the second element in the list, or the number 11 in this case, alright?

The sort function

Finally, let's have a built-in sort function that you can use:

```
z = [3, 2, 1]
z.sort()
z
```

So if I start with list `z`, which is 3, 2, and 1, I can call sort on that list, and `z` will now be sorted in order. The output of the above code is as follows:

```
[1, 2, 3]
```

Reverse sort

```
z.sort(reverse=True)
z
```

The output of the above code is as follows:

```
[3, 2, 1]
```

If you need to do a reverse sort, you can just say `reverse=True` as an attribute, as a parameter in that `sort` function, and that will put it back to 3, 2, 1.

If you need to let that sink in a little bit, feel free to go back and read it a little bit more.

Tuples

Tuples are just like lists, except they're immutable, so you can't actually extend, append, or sort them. They are what they are, and they behave just like lists, apart from the fact that you can't change them, and you indicate that they are immutable and are tuple, as opposed to a list, using parentheses instead of a square bracket. So you can see they work pretty much the same way otherwise:

```
#Tuples are just immutable lists. Use () instead of []
x = (1, 2, 3)
len(x)
```

Chapter 1

The output of the previous code is as follows:

```
3
```

We can say x= (1, 2, 3). I can still use length - len on that to say that there are three elements in that tuple, and even though, if you're not familiar with the term tuple, a tuple can actually contain as many elements as you want. Even though it sounds like it's Latin based on the number three, it doesn't mean you have three things in it. Usually, it only has two things in it. They can have as many as you want, really.

Dereferencing an element

We can also dereference the elements of a tuple, so element number 2 again would be the third element, because we start counting from 0, and that will give me back the number 6 in the following screenshot:

```
y = (4, 5, 6)
y[2]
```

The output to the above code is as follows:

```
6
```

List of tuples

We can also, like we could with lists, use tuples as elements of a list.

```
listOfTuples = [x, y]
listOfTuples
```

The output to the above code is as follows:

```
[(1, 2, 3), (4, 5, 6)]
```

Getting Started

We can create a new list that contains two tuples. So in the preceding example, we have our x tuple of (1, 2, 3) and our y tuple of (4, 5, 6); then we make a list of those two tuples and we get back this structure, where we have square brackets indicating a list that contains two tuples indicated by parentheses, and one thing that tuples are commonly used for when we're doing data science or any sort of managing or processing of data really is to use it to assign variables to input data as it's read in. I want to walk you through a little bit on what's going on in the following example:

```
(age, income) = "32,120000".split(',')
print (age)
print (income)
```

The output to the above code is as follows:

```
32
120000
```

Let's say we have a line of input data coming in and it's a comma-separated value file, which contains ages, say 32, comma-delimited by an income, say 120000 for that age, just to make something up. What I can do is as each line comes in, I can call the split function on it to actually separate that into a pair of values that are delimited by commas, and take that resulting tuple that comes out of split and assign it to two variables-age and income-all at once by defining a tuple of age, income and saying that I want to set that equal to the tuple that comes out of the split function.

So this is basically a common shorthand you'll see for assigning multiple fields to multiple variables at once. If I run that, you can see that the age variable actually ends up assigned to 32 and income to 120,000 because of that little trick there. You do need to be careful when you're doing this sort of thing, because if you don't have the expected number of fields or the expected number of elements in the resulting tuple, you will get an exception if you try to assign more stuff or less stuff than you expect to see here.

Dictionaries

Finally, the last data structure that we'll see a lot in Python is a dictionary, and you can think of that as a map or a hash table in other languages. It's a way to basically have a sort of mini-database, sort of a key/value data store that's built into Python. So let's say, I want to build up a little dictionary of Star Trek ships and their captains:

```
In [8]: # Like a map or hash table in other languages
        captains = {}
        captains["Enterprise"] = "Kirk"
        captains["Enterprise D"] = "Picard"
        captains["Deep Space Nine"] = "Sisko"
        captains["Voyager"] = "Janeway"

        print (captains['Voyager'])

        Janeway

In [9]: print (captains.get("Enterprise"))

        Kirk

In [10]: print (captains.get("NX-01"))

        None

In [11]: for ship in captains:
             print (ship + ":" + captains[ship])

        Deep Space Nine:Sisko
        Enterprise:Kirk
        Voyager:Janeway
        Enterprise D:Picard
```

I can set up a `captains = {}`, where curly brackets indicates an empty dictionary. Now I can use this sort of a syntax to assign entries in my dictionary, so I can say `captains` for `Enterprise` is `Kirk`, for `Enterprise D` it is `Picard`, for `Deep Space Nine` it is `Sisko`, and for `Voyager` it is `Janeway`. Now I have, basically, this lookup table that will associate ship names with their captain, and I can say, for example, `print captains["Voyager"]`, and I get back `Janeway`.

A very useful tool for basically doing lookups of some sort. Let's say you have some sort of an identifier in a dataset that maps to some human-readable name. You'll probably be using a dictionary to actually do that look up when you're printing it out.

We can also see what happens if you try to look up something that doesn't exist. Well, we can use the `get` function on a dictionary to safely return an entry. So in this case, `Enterprise` does have an entry in my dictionary, it just gives me back `Kirk`, but if I call the `NX-01` ship on the dictionary, I never defined the captain of that, so it comes back with a `None` value in this example, which is better than throwing an exception, but you do need to be aware that this is a possibility:

```
print (captains.get("NX-01"))
```

The output of the above code is as follows:

```
None
```

The captain is Jonathan Archer, but you know, I'm get a little bit too geeky here now.

Iterating through entries

```
for ship in captains:
    print (ship + ": " + captains[ship])
```

The output of the above code is as follows:

```
Enterprise D: Picard
Deep Space Nine: Sisko
Enterprise: Kirk
Voyager: Janeway
```

Let's look at a little example of iterating through the entries in a dictionary. If I want to iterate through every ship that I have in my dictionary and print out `captains`, I can type for `ship` in `captains`, and this will iterate through every single key in my dictionary. Then I can print out the lookup value of each ship's captain, and that's the output that I get there.

There you have it. This is basically the main data structures that you'll encounter in Python. There are some others, such as sets, but we'll not really use them in this book, so I think that's enough to get you started. Let's dive into some more Python nuances in our next section.

Python basics - Part 2

In addition to *Python Basics - Part 1*, let us now try to grasp more Python concepts in detail.

Functions in Python

Let's talk about functions in Python. Like with other languages, you can have functions that let you repeat a set of operations over and over again with different parameters. In Python, the syntax for doing that looks like this:

```
def SquareIt(x):
    return x * x
print (SquareIt(2))
```

The output of the above code is as follows:

```
4
```

You declare a function using the `def` keyword. It just says this is a function, and we'll call this function `SquareIt`, and the parameter list is then followed inside parentheses. This particular function only takes one parameter that we'll call x. Again, remember that whitespace is important in Python. There's not going to be any curly brackets or anything enclosing this function. It's strictly defined by whitespace. So we have a colon that says that this function declaration line is over, but then it's the fact that it's tabbed by one or more tabs that tells the interpreter that we are in fact within the `SquareIt` function.

So `def SquareIt(x):` tab returns x * x, and that will return the square of x in this function. We can go ahead and give that a try. `print squareIt(2)` is how we call that function. It looks just like it would be in any other language, really. This should return the number 4; we run the code, and in fact it does. Awesome! That's pretty simple, that's all there is to functions. Obviously, I could have more than one parameter if I wanted to, even as many parameters as I need.

Now there are some weird things you can do with functions in Python, that are kind of cool. One thing you can do is to pass functions around as though they were parameters. Let's take a closer look at this example:

```
#You can pass functions around as parameters
def DoSomething(f, x):
    return f(x)
print (DoSomething(SquareIt, 3))
```

Getting Started

The output of the preceding code is as follows:

```
9
```

Now I have a function called `DoSomething`, `def DoSomething`, and it will take two parameters, one that I'll call `f` and the other I'll call `x`, and if I happen, I can actually pass in a function for one of these parameters. So, think about that for a minute. Look at this example with a bit more sense. Here, `DoSomething(f, x):` will return `f` of `x`; it will basically call the f function with x as a parameter, and there's no strong typing in Python, so we have to just kind of make sure that what we are passing in for that first parameter is in fact a function for this to work properly.

For example, we'll say print `DoSomething`, and for the first parameter, we'll pass in `SquareIt`, which is actually another function, and the number 3. What this should do is to say do something with the `SquareIt` function and the 3 parameter, and that will return `(SquareIt, 3)`, and 3 squared last time I checked was 9, and sure enough, that does in fact work.

This might be a little bit of a new concept to you, passing functions around as parameters, so if you need to stop for a minute there, wait and let that sink in, play around with it, please feel free to do so. Again, I encourage you to stop and take this at your own pace.

Lambda functions - functional programming

One more thing that's kind of a Python-ish sort of a thing to do, which you might not see in other languages is the concept of lambda functions, and it's kind of called **functional programming**. The idea is that you can include a simple function into a function. This makes the most sense with an example:

```
#Lambda functions let you inline simple functions
print (DoSomething(lambda x: x * x * x, 3))
```

The output of the above code is as follows:

```
27
```

We'll print `DoSomething`, and remember that our first parameter is a function, so instead of passing in a named function, I can declare this function inline using the `lambda` keyword. Lambda basically means that I'm defining an unnamed function that just exists for now. It's transitory, and it takes a parameter x. In the syntax here, `lambda` means I'm defining an inline function of some sort, followed by its parameter list. It has a single parameter, `x`, and the colon, followed by what that function actually does. I'll take the x parameter and multiply it by itself three times to basically get the cube of a parameter.

In this example, `DoSomething` will pass in this lambda function as the first parameter, which computes the cube of x and the 3 parameter. So what's this really doing under the hood? This `lambda` function is a function of itself that gets passed into the f in `DoSomething` in the previous example, and x here is going to be 3. This will return f of x, which will end up executing our lambda function on the value 3. So that 3 goes into our x parameter, and our lambda function transforms that into 3 times 3 times 3, which is, of course, 27.

Now this comes up a lot when we start doing MapReduce and Spark and things like that. So if we'll be dealing with Hadoop sorts of technologies later on, this is a very important concept to understand. Again, I encourage you to take a moment to let that sink in and understand what's going on there if you need to.

Understanding boolean expressions

Boolean expression syntax is a little bit weird or unusual, at least in Python:

```
print (1 == 3)
```

The output of the above code is as follows:

```
False
```

As usual, we have the double equal symbol that can test for equality between two values. So does 1 equal 3, no it doesn't, therefore `False`. The value `False` is a special value designated by F. Remember that when you're trying to test, when you're doing Boolean stuff, the relevant keywords are `True` with a T and `False` with an F. That's a little bit different from other languages that I've worked with, so keep that in mind.

```
print (True or False)
```

The output of the above code is as follows:

```
True
```

Well, `True` or `False` is `True`, because one of them is `True`, you run it and it comes back `True`.

The if statement

```
print (1 is 3)
```

Getting Started

The output of the previous code is as follows:

```
False
```

The other thing we can do is use `is`, which is sort of the same thing as equal. It's a more Python-ish representation of equality, so `1 == 3` is the same thing as `1 is 3`, but this is considered the more Pythonic way of doing it. So `1 is 3` comes back as `False` because 1 is not 3.

The if-else loop

```
if 1 is 3:
    print "How did that happen?"
elif 1 > 3:
    print ("Yikes")
else:
    print ("All is well with the world")
```

The output of the above code is as follows:

```
All is well with the world
```

We can also do `if-else` and `else-if` blocks here too. Let's do something a little bit more complicated here. If `1 is 3`, I would print `How did that happen?` But of course 1 is not 3, so we will fall back down to the `else-if` block, otherwise, if 1 is not 3, we'll test if `1 > 3`. Well that's not true either, but if it did, we print `Yikes`, and we will finally fall into this catch-all `else` clause that will print `All is well with the world`.

In fact, 1 is not 3, nor is 1 greater than 3, and sure enough, `All is well with the world`. So, you know, other languages have very similar syntax, but these are the peculiarities of Python and how to do an `if-else` or `else-if` block. So again, feel free to keep this notebook around. It might be a good reference later on.

Looping

The last concept I want to cover in our Python basics is looping, and we saw a couple of examples of this already, but let's just do another one:

```
for x in range(10):
    print (x),
```

The output of the previous code is as follows:

```
0 1 2 3 4 5 6 7 8 9
```

For example, we can use this range operator to automatically define a list of numbers in the range. So if we say `for x in range(10)`, `range 10` will produce a list of 0 through 9, and by saying for x in that list, we will iterate through every individual entry in that list and print it out. Again, the comma after the `print` statement says don't give me a new line, just keep on going. So the output of this ends up being all the elements of that list printed next to each other.

To do something a little bit more complicated, we'll do something similar, but this time we'll show how `continue` and `break` work. As in other languages, you can actually choose to skip the rest of the processing for a loop iteration, or actually stop the iteration of the loop prematurely:

```
for x in range(10):
    if (x is 1):
   continue
   if (x > 5):
      break
   print (x),
```

The output of the above code is as follows:

```
0 2 3 4 5
```

In this example, we'll go through the values 0 through 9, and if we hit on the number 1, we will continue before we print it out. We'll skip the number 1, basically, and if the number is greater than 5, we'll break the loop and stop the processing entirely. The output that we expect is that we will print out the numbers 0 through 5, unless it's 1, in which case, we'll skip number 1, and sure enough, that's what it does.

The while loop

Another syntax is the while loop. This is kind of a standard looping syntax that you see in most languages:

```
x = 0
while (x < 10):
    print (x),
    x += 1
```

The output of the previous code is as follows:

```
0 1 2 3 4 5 6 7 8 9
```

We can also say, start with `x = 0`, and `while (x < 10) :`, print it out and then increment x by 1. This will go through over and over again, incrementing x until it's less than 10, at which point we break out of the `while` loop and we're done. So it does the same thing as this first example here, but just in a different style. It prints out the numbers 0 through 9 using a `while` loop. Just some examples there, nothing too complicated. Again, if you've done any sort of programming or scripting before, this should be pretty simple.

Now to really let this sink in, I've been saying throughout this entire chapter, get in there, get your hands dirty, and play with it. So I'm going to make you do that.

Exploring activity

Here's an activity, a little bit of a challenge for you:

Here's a nice little code block where you can start writing your own Python code, run it, and play around with it, so please do so. Your challenge is to write some code that creates a list of integers, loops through each element of that list, pretty easy so far, and only prints out even numbers.

Now this shouldn't be too hard. There are examples in this notebook of doing all that stuff; all you have to do is put it together and get it to run. So, the point is not to give you something that's hard. I just want you to actually get some confidence in writing your own Python code and actually running it and seeing it operate, so please do so. I definitely encourage you to be interactive here. So have at it, good luck, and welcome to Python.

So that's your Python crash course, obviously, just some very basic stuff there. As we go through more and more examples throughout the book, it'll make more and more sense since you have more examples to look at, but if you do feel a little bit intimidated at this point, maybe you're a little bit too new to programming or scripting, and it might be a good idea to go and take a Python revision before moving forward, but if you feel pretty good about what you've seen so far, let's move ahead and we'll keep on going.

Running Python scripts

Throughout this book, we'll be using the IPython/Jupyter Notebook format (which are `.ipynb` files) that we've been looking at so far, and it's a great format for a book like this because it lets me put little blocks of code in there and put a little text and things around it explaining what it's doing, and you can experiment with things live.

Of course, it's great from that standpoint, but in the real world, you're probably not going to be using IPython/Jupyter Notebooks to actually run your Python scripts in production, so let me just really briefly go through the other ways you can run Python code, and other interactive ways of running Python code as well. So it's a pretty flexible system. Let's take a look.

Getting Started

More options than just the IPython/Jupyter Notebook

I want to make sure that you know there's more than one way to run Python code. Now, throughout this book, we'll be using the IPython/Jupyter Notebook format but in the real world, you're not going to be running your code as a notebook. You're going to be running it as a standalone Python script. So I just want to make sure you know how to do that and see how it works.

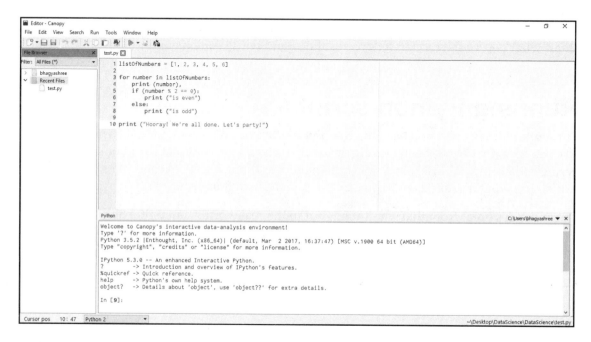

So let's go back to this first example that we ran in the book, just to illustrate the importance of whitespace. We can just select and copy that code out of the notebook format and paste it into a new file.

This can be done by clicking on the New button at the extreme left. So let's make a new file and paste it in and let's save this file and call it, `test.py`, where `py` is the usual extension that we give to Python scripts. Now, I can run this in a few different ways.

Running Python scripts in command prompt

I can actually run the script in a command prompt. If I go to **Tools**, I can go to **Canopy Command Prompt**, and that will open up a command window that has all the necessary environment variables already in place for running Python. I can just type `python test.py` and run the script, and out comes my result:

```
(Canopy 64bit) E:\DataScience>python test.py
1 is odd
2 is even
3 is odd
4 is even
5 is odd
6 is even
Hooray! We're all done. Let's party!

(Canopy 64bit) E:\DataScience>
```

So in the real world, you'd probably do something like that. It might be on a Crontab or something like that, who knows? But running a real script in production is just that simple. You can now close the command prompt.

Getting Started

Using the Canopy IDE

Moving back, I can also run the script from within the IDE. So from within Canopy, I can go to the **Run** menu. I can either go to **Run** | **Run File**, or click on the little play icon, and that will also execute my script, and see the results at the bottom in the output window, as shown in the following screenshot:

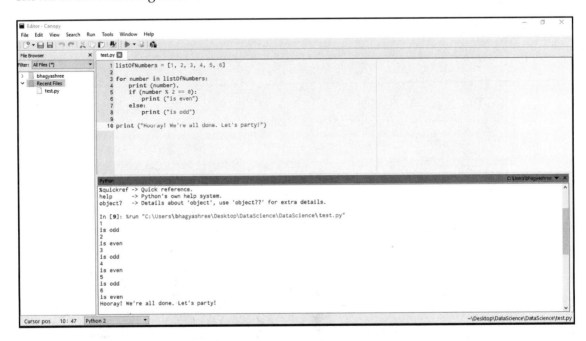

So that's another way to do it, and finally, you can also run scripts within this interactive prompt present at the bottom interactively. I can actually type in Python commands one at a time down, and have them just execute and stay within the environment down there:

For example, I could say `stuff`, make it a `list` call, and have 1, 2, 3, 4, and now I can say `len(stuff)`, and that will give me 4:

```
In [9]: stuff= [1, 2, 3, 4]

In [10]: len(stuff)
Out[10]: 4
```

I can say, `for x in stuff:print x`, and we get output as 1 2 3 4:

```
In [7]: for x in stuff:
   ...:     print (x),
   ...:
1
2
3
4

In [8]:
```

So you can see you can kind of makeup scripts as you go down in the interactive prompt at the bottom and execute things one thing at a time. In this example, `stuff` is a variable we created, a list that stays in memory, it's kind of like a global variable in other languages within this environment.

Now if I do want to reset this environment, if I want to get rid of `stuff` and start all over, the way you do that is you go up to the **Run** menu here and you can say **Restart Kernel**, and that will strike you over with a blank slate:

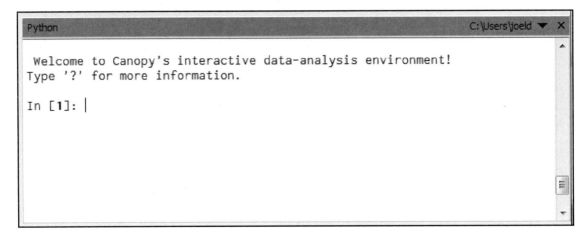

So now I have a new Python environment that's a clean slate, and in this case, what did I call it? Type `stuff` and `stuff` doesn't exist yet because I have a new environment, but I can make it something else, such as `[4, 5, 6]`; run it and there it is:

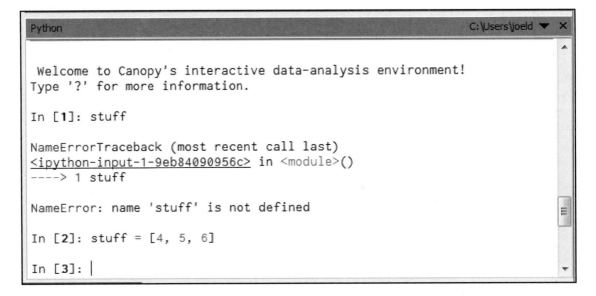

So there you have it, three ways of running Python code: the IPython/Jupyter Notebook, which we'll use throughout this book just because it's a good learning tool, you can also run scripts as standalone script files, and you can also execute Python code in the interactive command prompt.

So there you have it, and there you have three different ways of running Python code and experimenting and running things in production. So keep that in mind. We'll be using notebooks throughout the rest of this book, but again, you have those other options when the time comes.

Summary

In this chapter, we started our journey with building the most important stepping stone of the book - Installing Enthought Canopy. We then moved to installing other libraries and installing different types of packages. We also grasped some of the basics of Python with the help of various Python code. We covered basic concepts such as modules, lists along with Tuples, and eventually moved on to understanding more of Python basics with a better knowledge of functions and looping in Python. Finally, we started with running some of our simple Python scripts.

In the next chapter, we will move on to understand concepts of statistics and probability.

2
Statistics and Probability Refresher, and Python Practice

In this chapter, we are going to go through a few concepts of statistics and probability, which might be a refresher for some of you. These concepts are important to go through if you want to be a data scientist. We will see examples to understand these concepts better. We will also look at how to implement those examples using actual Python code.

We'll be covering the following topics in this chapter:

- Types of data you may encounter and how to treat them accordingly
- Statistical concepts of mean, median, mode, standard deviation, and variance
- Probability density functions and probability mass functions
- Types of data distributions and how to plot them
- Understanding percentiles and moments

Types of data

Alright, if you want to be a data scientist, we need to talk about the types of data that you might encounter, how to categorize them, and how you might treat them differently. Let's dive into the different flavors of data you might encounter:

This will seem pretty basic, but we've got to start with the simple stuff and we'll work our way up to the more complicated data mining and machine learning things. It is important to know what kind of data you're dealing with because different techniques might have different nuances depending on what kind of data you're handling. So, there are several flavors of data, if you will, and there are three specific types of data that we will primarily focus on. They are:

- Numerical data
- Categorical data
- Ordinal data

Again, there are different variations of techniques that you might use for different types of data, so you always need to keep in mind what kind of data you're dealing with when you're analyzing it.

Numerical data

Let's start with numerical data. It's probably the most common data type. Basically, it represents some quantifiable thing that you can measure. Some examples are heights of people, page load times, stock prices, and so on. Things that vary, things that you can measure, things that have a wide range of possibilities. Now there are basically two kinds of numerical data, so a flavor of a flavor if you will.

Discrete data

There's discrete data, which is integer-based and, for example, can be counts of some sort of event. Some examples are how many purchases did a customer make in a year. Well, that can only be discrete values. They bought one thing, or they bought two things, or they bought three things. They couldn't have bought, 2.25 things or three and three-quarters things. It's a discrete value that has an integer restriction to it.

Continuous data

The other type of numerical data is continuous data, and this is stuff that has an infinite range of possibilities where you can go into fractions. So, for example, going back to the height of people, there is an infinite number of possible heights for people. You could be five feet and 10.37625 inches tall, or the time it takes to do something like check out on a website could be any huge range of possibilities, 10.7625 seconds for all you know, or how much rainfall in a given day. Again, there's an infinite amount of precision there. So that's an example of continuous data.

To recap, numerical data is something you can measure quantitatively with a number, and it can be either discrete, where it's integer-based like an event count, or continuous, where you can have an infinite range of precision available to that data.

Categorical data

The second type of data that we're going to talk about is categorical data, and this is data that has no inherent numeric meaning.

Most of the time, you can't really compare one category to another directly. Things like gender, yes/no questions, race, state of residence, product category, political party; you can assign numbers to these categories, and often you will, but those numbers have no inherent meaning.

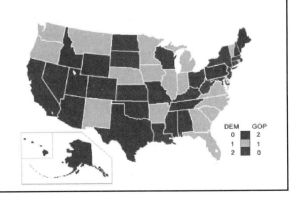

So, for example, I can say that the area of Texas is greater than the area of Florida, but I can't just say Texas is greater than Florida, they're just categories. There's no real numerical quantifiable meaning to them, it's just ways that we categorize different things.

Now again, I might have some sort of numerical assignment to each state. I mean, I could say that Florida is state number 3 and Texas state number 4, but there's no real relationship between 3 and 4 there, right, it's just a shorthand to more compactly represent these categories. So again, categorical data does not have any intrinsic numerical meaning; it's just a way that you're choosing to split up a set of data based on categories.

Chapter 2

Ordinal data

The last category that you tend to hear about with types of data is ordinal data, and it's sort of a mixture of numerical and categorical data. A common example is star ratings for a movie or music, or what have you.

Ordinal

- A mixture of numerical and categorical
- Categorical data that has mathematical meaning
- Example: movie ratings on a 1-5 scale.
  - Ratings must be 1, 2, 3, 4, or 5
  - But these values have mathematical meaning; 1 means it's a worse movie than a 2.

In this case, we have categorical data in that could be 1 through 5 stars, where 1 might represent poor and 5 might represent excellent, but they do have mathematical meaning. We do know that 5 means it's better than a 1, so this is a case where we have data where the different categories have a numerical relationship to each other. So, I can say that 1 star is less than 5 stars, I can say that 2 stars is less than 3 stars, I can say that 4 stars is greater than 2 stars in terms of a measure of quality. Now you could also think of the actual number of stars as discrete numerical data. So, it's definitely a fine line between these categories, and in a lot of cases you can actually treat them interchangeably.

So, there you have it, the three different types. There is numerical, categorical, and ordinal data. Let's see if it's sunk in. Don't worry, I'm not going to make you hand in your work or anything.

Quick quiz: For each of these examples, is the data numerical, categorical, or ordinal?

1. Let's start with how much gas is in your gas tank. What do you think? Well, the right answer is numerical. It's a continuous numerical value because you can have an infinite range of possibilities of gas in your tank. I mean, yeah, there's probably some upper bound of how much gas you can fit in it, but there is no end to the number of possible values of how much gas you have. It could be three quarters of a tank, it could be seven sixteenths of the tank, it could be *1/pi* of a tank, I mean who knows, right?

2. How about if you're reading your overall health on a scale of 1 to 4, where those choices correspond to the categories poor, moderate, good, and excellent? What do you think? That's a good example of ordinal data. That's very much like our movie ratings data, and again, depending on how you model that, you could probably treat it as discrete numerical data as well, but technically we're going to call that ordinal data.
3. What about the races of your classmates? This is a pretty clear example of categorical data. You can't really compare purple people to green people, right, they're just purple and green, but they are categories that you might want to study and understand the differences between on some other dimension.
4. How about the ages of your classmates in years? A little bit of a trick question there; if I said it had to be in an integer value of years, like 40, 50, or 55 years old, then that would be discrete numerical data, but if I had more precision, like 40 years three months and 2.67 days, that would be continuous numerical data, but either way, it's a numerical data type.
5. And finally, money spent in a store. Again, that could be an example of continuous numerical data. So again, this is only important because you might apply different techniques to different types of data.

There might be some concepts where we do one type of implementation for categorical data and a different type of implementation for numerical data, for example.

So that's all you need to know about the different types of data that you'll commonly find, and that we'll focus on in this book. They're all pretty simple concepts: you've got numeric, categorical, and ordinal data, and numerical data can be continuous or discrete. There might be different techniques you apply to the data depending on what kind of data you're dealing with, and we'll see that throughout the book. Let's move on.

Mean, median, and mode

Let's do a little refresher of statistics 101. This is like elementary school stuff, but good to go through it again and see how these different techniques are used: Mean, median, and mode. I'm sure you've heard those terms before, but it's good to see how they're used differently, so let's dive in.

This should be a review for most of you, a quick refresher, now that we're starting to actually dive into some real statistics. Let's look at some actual data and figure out how to measure these things.

Mean

The mean, as you probably know, is just another name for the average. To calculate the mean of a dataset, all you have to do is sum up all the values and divide it by the number of values that you have.

Sum of samples/Number of samples

Let's take this example, which calculates the mean (average) number of children per house in my neighborhood.

Let's say I went door-to-door in my neighborhood and asked everyone, how many children live in their household. (That, by the way, is a good example of discrete numerical data; remember from the previous section?) Let's say I go around and I found out that the first house has no kids in it, and the second house has two children, and the third household has three children, and so on and so forth. I amassed this little dataset of discrete numerical data, and to figure out the mean, all I do is add them all together and divide it by the number of houses that I went to.

Number of children in each house on my street:

0, 2, 3, 2, 1, 0, 0, 2, 0

The mean is *(0+2+3+2+1+0+0+2+0)/9 = 1.11*

It comes out as 0 plus 2 plus 3 plus all the rest of these numbers divided by the total number of houses that I looked at, which is 9, and the mean number of children per house in my sample is 1.11. So, there you have it, mean.

Median

Median is a little bit different. The way you compute the median of the dataset is by sorting all the values (in either ascending or descending order), and taking the one that ends up in the middle.

So, for example, let's use the same dataset of children in my neighborhood

0, 2, 3, 2, 1, 0, 0, 2, 0

I would sort it numerically, and I can take the number that's slap dab in the middle of the data, which turns out to be 1.

0, 0, 0, 0, 1, 2, 2, 2, 3

Again, all I do is take the data, sort it numerically, and take the center point.

If you have an even number of data points, then the median might actually fall in between two data points. It wouldn't be clear which one is actually the middle. In that case, all you do is, take the average of the two that do fall in the middle and consider that number as the median.

The factor of outliers

Now in the preceding example of the number of kids in each household, the median and the mean were pretty close to each other because there weren't a lot of outliers. We had 0, 1, 2, or 3 kids, but we didn't have some wacky family that had 100 kids. That would have really skewed the mean, but it might not have changed the median too much. That's why the median is often a very useful thing to look at and often overlooked.

Median is less susceptible to outliers than the mean.

People have a tendency to mislead people with statistics sometimes. I'm going to keep pointing this out throughout the book wherever I can.

For example, you can talk about the mean or average household income in the United States, and that actual number from last year when I looked it up was $72,000 or so, but that doesn't really provide an accurate picture of what the typical American makes. That is because, if you look at the median income, it's much lower at $51,939. Why is that? Well, because of income inequality. There are a few very rich people in America, and the same is true in a lot of countries as well. America's not even the worst, but you know those billionaires, those super-rich people that live on Wall Street or Silicon Valley or some other super-rich place, they skew the mean. But there's so few of them that they don't really affect the median so much.

This is a great example of where the median tells a much better story about the typical person or data point in this example than the mean does. Whenever someone talks about the mean, you have to think about what does the data distribution looks like. Are there outliers that might be skewing that mean? And if the answer is potentially yes, you should also ask for the median, because often, that provides more insight than the mean or the average.

Mode

Finally, we'll talk about mode. This doesn't really come up too often in practice, but you can't talk about mean and median without talking about mode. All mode means, is the most common value in a dataset.

Let's go back to my example of the number of kids in each house.

0, 2, 3, 2, 1, 0, 0, 2, 0

How many of each value are there:

0: 4, 1: 1, 2: 3, 3: 1

The MODE is 0

If I just look at what number occurs most frequently, it turns out to be 0, and the mode therefore of this data is 0. The most common number of children in a given house in this neighborhood is no kids, and that's all that means.

Now this is actually a pretty good example of continuous versus discrete data, because this only really works with discrete data. If I have a continuous range of data then I can't really talk about the most common value that occurs, unless I quantize that somehow into discrete values. So we've already run into one example here where the data type matters.

Mode is usually only relevant to discrete numerical data, and not to continuous data.

A lot of real-world data tends to be continuous, so maybe that's why I don't hear too much about mode, but we see it here for completeness.

There you have it: mean, median, and mode in a nutshell. Kind of the most basic statistics stuff you can possibly do, but I hope you gained a little refresher there in the importance of choosing between median and mean. They can tell very different stories, and yet people tend to equate them in their heads, so make sure you're being a responsible data scientist and representing data in a way that conveys the meaning you're trying to represent. If you're trying to display a typical value, often the median is a better choice than the mean because of outliers, so remember that. Let's move on.

Using mean, median, and mode in Python

Let's start doing some real coding in Python and see how you compute the mean, median, and mode using Python in an IPython Notebook file.

So go ahead and open up the `MeanMedianMode.ipynb` file from the data files for this section if you'd like to follow along, which I definitely encourage you to do. If you need to go back to that earlier section on where to download these materials from, please go do that, because you will need these files for the section. Let's dive in!

Calculating mean using the NumPy package

What we're going to do is create some fake income data, getting back to our example from the previous section. We're going to create some fake data where the typical American makes around $27,000 a year in this example, we're going to say that's distributed with a normal distribution and a standard deviation of 15,000. All numbers are completely made up, and if you don't know what normal distribution and standard deviation means yet, don't worry. I'm going to cover that a little later in the chapter, but I just want you to know what these different parameters represent in this example. It will make sense later on.

In our Python notebook, remember to import the NumPy package into Python, which makes computing mean, median, and mode really easy. We're going to use the `import numpy as np` directive, which means we can use `np` as a shorthand to call `numpy` from now on.

Then we're going to create a list of numbers called `incomes` using the `np.random.normal` function.

```
import numpy as np

incomes = np.random.normal(27000, 15000, 10000)
np.mean(incomes)
```

The three parameters of the `np.random.normal` function mean I want the data centered around `27000`, with a standard deviation of `15000`, and I want python to make `10000` data points in this list.

Once I do that, I compute the average of those data points, or the mean by just calling `np.mean` on `incomes` which is my list of data. It's just that simple.

Let's go ahead and run that. Make sure you selected that code block and then you can hit the play button to actually execute it, and since there is a random component to these income numbers, every time I run it, I'm going to get a slightly different result, but it should always be pretty close to `27000`.

```
Out[1]: 27173.098561362742
```

OK, so that's all there is to computing the mean in Python, just using NumPy (`np.mean`) makes it super easy. You don't have to write a bunch of code or actually add up everything and count up how many items you had and do the division. NumPy mean, does all that for you.

Visualizing data using matplotlib

Let's visualize this data to make it make a little more sense. So there's another package called `matplotlib`, and again we're going to talk about that a lot more in the future as well, but it's a package that lets me make pretty graphs in IPython Notebooks, so it's an easy way to visualize your data and see what's going on.

In this example, we are using `matplotlib` to create a histogram of our income data broken up into `50` different buckets. So basically, we're taking our continuous data and discretizing it, and then we can call show on `matplotlib.pyplot` to actually display this histogram in line. Refer to the following code:

```
%matplotlib inline
import matplotlib.pyplot as plt
plt.hist(incomes, 50)
plt.show()
```

Go ahead and select the code block and hit play. It will actually create a new graph for us as follows:

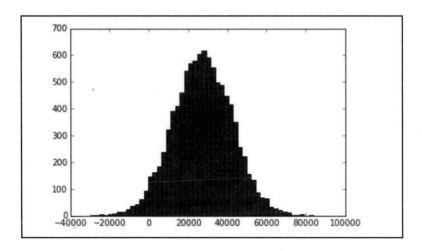

If you're not familiar with histograms or you need a refresher, the way to interpret this is that for each one of these buckets that we've discretized our data into is showing the frequency of that data.

So, for example, around 27,000-ish we see there's about **600** data points in that neighborhood for each given range of values. There's a lot of people around the 27,000 mark, but when you get over to outliers like **80,000**, there is not a whole lot, and apparently there's some poor souls that are even in debt at **-40,000**, but again, they're very rare and not probable because we defined a normal distribution, and this is what a normal probability curve looks like. Again, we're going to talk about that more in detail later, but I just want to get that idea in your head if you don't already know it.

Calculating median using the NumPy package

Alright, so computing the median is just as simple as computing the mean. Just like we had NumPy `mean`, we have a NumPy `median` function as well.

We can just use the `median` function on `incomes`, which is our list of data, and that will give us the median. In this case, that came up to $26,911, which isn't very different from the mean of $26988. Again, the initial data was random, so your values will be slightly different.

```
np.median(incomes)
```

The following is the output of the preceding code:

```
Out[4]: 26911.948365056276
```

We don't expect to see a lot of outliers because this is a nice normal distribution. Median and mean will be comparable when you don't have a lot of weird outliers.

Analyzing the effect of outliers

Just to prove a point, let's add in an outlier. We'll take Donald Trump; I think he qualifies as an outlier. Let's go ahead and add his income in. So I'm going to manually add this to the data using `np.append`, and let's say add a billion dollars (which is obviously not the actual income of Donald Trump) into the incomes data.

```
incomes = np.append(incomes, [1000000000])
```

What we're going to see is that this outlier doesn't really change the median a whole lot, you know, that's still going to be around the same value $26,911, because we didn't actually change where the middle point is, with that one value, as shown in the following example:

```
np.median(incomes)
```

This will output the following:

```
Out[5]: 26911.948365056276
```

This gives a new output of:

```
np.mean(incomes)
```

The following is the output of the preceding code:

```
Out[5]:127160.38252311043
```

Aha, so there you have it! It is a great example of how median and mean, although people tend to equate them in commonplace language, can be very different, and tell a very different story. So that one outlier caused the average income in this dataset to be over $127160 a year, but the more accurate picture is closer to 27,000 dollars a year for the typical person in this dataset. We just had the mean skewed by one big outlier.

The moral of the story is: take anyone who talks about means or averages with a grain of salt if you suspect there might be outliers involved, and income distribution is definitely a case of that.

Calculating mode using the SciPy package

Finally, let's look at mode. We will just generate a bunch of random integers, 500 of them to be precise, that range between `18` and `90`. We're going to create a bunch of fake ages for people.

```
ages = np.random.randint(18, high=90, size=500)
ages
```

Your output will be random, but should look something like the following screenshot:

```
Out[7]: array([69, 87, 31, 22, 78, 37, 77, 32, 18, 59, 29, 43, 34, 33, 56, 83, 66,
        30, 77, 74, 31, 21, 85, 50, 47, 26, 72, 62, 33, 45, 86, 50, 86, 56,
        31, 84, 78, 27, 76, 42, 83, 64, 48, 54, 70, 56, 24, 50, 50, 71, 49,
        20, 85, 61, 33, 83, 55, 21, 60, 80, 56, 89, 61, 56, 52, 55, 20, 31,
        69, 50, 21, 52, 31, 83, 43, 77, 27, 67, 39, 39, 26, 38, 40, 73, 50,
        31, 87, 23, 50, 34, 69, 45, 83, 51, 88, 41, 64, 59, 40, 89, 57, 62,
        55, 75, 38, 51, 24, 21, 18, 75, 58, 62, 81, 65, 89, 64, 43, 33, 53,
        72, 20, 56, 19, 26, 81, 68, 70, 70, 41, 59, 50, 77, 62, 31, 87, 58,
        63, 83, 35, 55, 38, 85, 53, 66, 28, 74, 42, 28, 80, 69, 54, 25, 74,
        58, 27, 42, 87, 46, 43, 44, 33, 40, 21, 21, 73, 48, 87, 63, 84, 55,
        61, 66, 48, 73, 27, 60, 34, 77, 59, 58, 50, 70, 30, 76, 72, 33, 80,
        43, 63, 49, 60, 61, 53, 55, 79, 38, 46, 38, 81, 66, 29, 81, 46, 19,
        49, 57, 31, 18, 25, 47, 20, 88, 33, 88, 50, 22, 57, 39, 20, 59, 63,
        38, 35, 59, 28, 23, 56, 50, 46, 65, 46, 88, 87, 34, 73, 75, 32, 49,
        67, 77, 86, 38, 80, 36, 64, 79, 65, 51, 46, 54, 23, 82, 56, 41, 78,
        19, 45, 38, 70, 74, 56, 87, 49, 69, 30, 25, 22, 71, 39, 41, 46, 72,
        33, 72, 88, 37, 75, 39, 37, 21, 67, 86, 77, 20, 46, 53, 22, 85, 73,
        89, 67, 24, 24, 25, 62, 56, 58, 44, 63, 30, 36, 73, 49, 45, 26, 33,
        20, 62, 75, 34, 81, 59, 64, 27, 43, 23, 62, 75, 81, 40, 65, 29, 61,
        55, 81, 35, 68, 79, 86, 43, 35, 74, 59, 80, 75, 60, 82, 66, 54, 37,
        54, 71, 88, 46, 55, 63, 79, 89, 48, 61, 68, 78, 51, 32, 26, 48, 78,
        76, 62, 19, 19, 63, 20, 44, 28, 34, 58, 44, 36, 70, 34, 67, 50, 33,
        31, 18, 72, 55, 49, 63, 81, 65, 51, 46, 22, 55, 77, 76, 53, 79, 47,
        57, 46, 27, 29, 49, 71, 19, 85, 86, 77, 89, 59, 67, 26, 50, 79, 85,
        68, 51, 30, 18, 73, 52, 22, 53, 56, 26, 45, 60, 83, 50, 34, 68, 65,
        27, 72, 24, 34, 37, 52, 67, 79, 79, 24, 65, 71, 28, 29, 61, 34, 77,
        35, 59, 50, 83, 27, 32, 18, 81, 36, 46, 48, 39, 52, 23, 37, 62, 54,
        53, 50, 34, 36, 88, 83, 39, 89, 65, 83, 73, 66, 28, 36, 56, 86, 65,
        28, 46, 18, 61, 69, 80, 85, 29, 85, 44, 18, 61, 68, 83, 89, 53, 65,
        55, 66, 87, 55, 43, 32, 84])
```

Now, SciPy, kind of like NumPy, is a bunch of like handy-dandy statistics functions, so we can import `stats` from SciPy using the following syntax. It's a little bit different than what we saw before.

```
from scipy import stats
stats.mode(ages)
```

The code means, from the `scipy` package import `stats`, and I'm just going to refer to the package as `stats`, Tha means that I don't need to have an alias like I did before with NumPy, just different way of doing it. Both ways work. Then, I used the `stats.mode` function on `ages`, which is our list of random ages. When we execute the above code, we get the following output:

```
Out[11]: ModeResult(mode=array([39]), count=array([12]))
```

So in this case, the actual mode is `39` that turned out to be the most common value in that array. It actually occurred `12` times.

Now if I actually create a new distribution, I would expect a completely different answer because this data really is completely random what these numbers are. Let's execute the above code blocks again to create a new distribution.

```
ages = np.random.randint(18, high=90, size=500)
ages

from scipy import stats
stats.mode(ages)
```

The output for randomizing the equation is as distribution is as follows:

```
Out[12]: array([41, 74, 26, 31, 31, 31, 20, 64, 59, 76, 80, 59, 53, 50, 29, 67, 55,
                82, 41, 40, 77, 41, 73, 52, 38, 87, 28, 87, 60, 47, 87, 66, 71, 77,
                85, 40, 22, 40, 74, 69, 44, 46, 72, 60, 69, 56, 19, 84, 80, 83, 22,
                63, 74, 31, 32, 20, 58, 71, 56, 43, 32, 67, 32, 51, 79, 54, 25, 81,
                50, 55, 86, 75, 30, 37, 37, 37, 56, 22, 85, 82, 58, 78, 32, 50, 52,
                70, 85, 37, 34, 83, 41, 52, 46, 55, 84, 64, 19, 86, 46, 65, 77, 80,
                82, 86, 65, 41, 35, 44, 45, 34, 46, 51, 83, 82, 53, 50, 84, 83, 29,
                47, 80, 75, 72, 81, 40, 75, 74, 57, 27, 71, 76, 65, 27, 75, 32, 26,
                34, 20, 58, 19, 18, 26, 73, 60, 31, 34, 46, 80, 76, 30, 70, 68, 71,
                45, 44, 47, 30, 39, 35, 60, 44, 45, 83, 64, 21, 35, 25, 70, 86, 53,
                65, 87, 66, 88, 48, 18, 29, 60, 50, 29, 67, 45, 83, 76, 62, 25, 41,
                56, 23, 60, 56, 59, 77, 64, 74, 39, 43, 24, 55, 88, 60, 19, 32, 49,
                59, 88, 69, 82, 56, 70, 34, 52, 85, 70, 79, 26, 37, 60, 40, 32, 20,
                81, 43, 47, 83, 67, 27, 30, 21, 24, 40, 43, 83, 79, 47, 36, 66, 37,
                76, 20, 48, 81, 58, 62, 27, 21, 88, 31, 62, 38, 83, 33, 41, 68, 38,
                43, 44, 49, 51, 82, 48, 53, 75, 56, 48, 38, 76, 37, 41, 62, 26, 32,
                53, 40, 89, 40, 19, 29, 73, 71, 81, 63, 36, 56, 30, 60, 67, 47, 20,
                62, 86, 84, 88, 37, 47, 35, 37, 26, 48, 36, 53, 19, 77, 46, 63, 87,
                60, 40, 72, 86, 41, 58, 29, 43, 36, 69, 75, 56, 55, 33, 66, 22, 46,
                73, 45, 30, 42, 51, 24, 18, 54, 45, 73, 37, 54, 84, 29, 73, 82, 47,
                55, 68, 42, 60, 25, 46, 89, 37, 20, 34, 24, 40, 61, 66, 72, 71, 30,
                50, 29, 24, 60, 30, 76, 67, 66, 19, 75, 33, 21, 21, 45, 38, 47, 69,
                71, 83, 50, 40, 24, 38, 47, 72, 25, 26, 77, 44, 39, 35, 36, 42, 73,
                78, 77, 62, 43, 84, 66, 41, 48, 69, 65, 52, 45, 85, 43, 77, 31, 50,
                61, 69, 71, 77, 89, 65, 41, 35, 88, 37, 87, 75, 21, 38, 73, 31, 66,
                25, 25, 69, 71, 46, 86, 66, 82, 24, 77, 44, 81, 72, 25, 50, 58, 22,
                85, 42, 44, 62, 71, 89, 77, 29, 65, 62, 62, 26, 65, 21, 49, 37, 82,
                26, 72, 26, 35, 45, 51, 63, 87, 25, 29, 72, 53, 33, 76, 65, 22, 22,
                87, 40, 46, 46, 89, 52, 55, 44, 66, 71, 78, 44, 70, 51, 73, 74, 44,
                71, 53, 84, 76, 61, 76, 33])
```

Make sure you selected that code block and then you can hit the play button to actually execute it.

In this case, the mode ended up being the number 29, which occurred 14 times.

```
Out[11]: ModeResult(mode=array([29]), count=array([14]))
```

So, it's a very simple concept. You can do it a few more times just for fun. It's kind of like rolling the roulette wheel. We'll create a new distribution again.

There you have it, mean, median, and mode in a nutshell. It's very simple to do using the SciPy and NumPy packages.

Some exercises

I'm going to give you a little assignment in this section. If you open up `MeanMedianExercise.ipynb` file, there's some stuff you can play with. I want you to roll up your sleeves and actually try to do this.

In the file, we have some random e-commerce data. What this data represents is the total amount spent per transaction, and again, just like with our previous example, it's just a normal distribution of data. We can run that, and your homework is to go ahead and find the mean and median of this data using the NumPy package. Pretty much the easiest assignment you could possibly imagine. All the techniques you need are in the `MeanMedianMode.ipynb` file that we used earlier.

The point here is not really to challenge you, it's just to make you actually write some Python code and convince yourself that you can actually get a result and make something happen here. So, go ahead and play with that. If you want to play with it some more, feel free to play around with the data distribution here and see what effect you can have on the numbers. Try adding some outliers, kind of like we did with the income data. This is the way to learn this stuff: master the basics and the advance stuff will follow. Have at it, have fun.

Once your're ready, let's move forward to our next concept, standard deviation and variance.

Standard deviation and variance

Let's talk about standard deviation and variance. The concepts and terms you've probably heard before, but let's go into a little bit more depth about what they really mean and how you compute them. It's a measure of the spread of a data distribution, and that will make a little bit more sense in a few minutes.

Standard deviation and variance are two fundamental quantities for a data distribution that you'll see over and over again in this book. So, let's see what they are, if you need a refresher.

Variance

Let's look at a histogram, because variance and standard deviation are all about the spread of the data, the shape of the distribution of a dataset. Take a look at the following histogram:

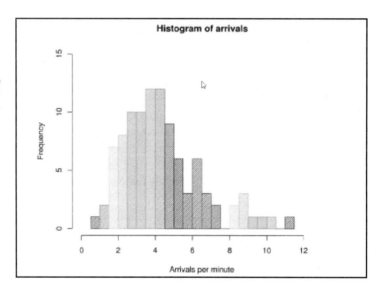

Let's say that we have some data on the arrival frequency of airplanes at an airport, for example, and this histogram indicates that we have around 4 arrivals per minute and that happened on around 12 days that we looked at for this data. However, we also have these outliers. We had one really slow day that only had one arrival per minute, we only had one really fast day where we had almost 12 arrivals per minute. So, the way to read a histogram is look up the bucket of a given value, and that tells you how frequently that value occurred in your data, and the shape of the histogram could tell you a lot about the probability distribution of a given set of data.

We know from this data that our airport is very likely to have around 4 arrivals per minute, but it's very unlikely to have 1 or 12, and we can also talk specifically about the probabilities of all the numbers in between. So not only is it unlikely to have 12 arrivals per minute, it's also very unlikely to have 9 arrivals per minute, but once we start getting around 8 or so, things start to pick up a little bit. A lot of information can be had from a histogram.

 Variance measures how *spread-out* the data is.

Chapter 2

Measuring variance

We usually refer to variance as sigma squared, and you'll find out why momentarily, but for now, just know that variance is the average of the squared differences from the mean.

1. To compute the variance of a dataset, you first figure out the mean of it. Let's say I have some data that could represent anything. Let's say maximum number of people that were standing in line for a given hour. In the first hour, I observed 1 person standing in line, then 4, then 5, then 4, then 8.
2. The first step in computing the variance is just to find the mean, or the average, of that data. I add them all, divide the sum by the number of data points, and that comes out to 4.4 which is the average number of people standing in line (1+4+5+4+8)/5 = 4.4.
3. Now the next step is to find the differences from the mean for each data point. I know that the mean is 4.4. So for my first data point, I have 1, so 1 - 4.4 = -3.4, The next data point is 4, so 4 - 4.4 = -0.4 4 - 4.4 = -0.4, and so on and so forth. OK, so I end up with these both positive and negative numbers that represent the variance from the mean for each data point *(-3.4, -0.4, 0.6, -0.4, 3.6)*.
4. Now what I need is a single number that represents the variance of this entire dataset. So, the next thing I'm going to do is find the square of these differences. I'm just going to go through each one of those raw differences from the mean and square them. This is for a couple of different reasons:

    - First, I want to make sure that negative variances. Count just as much as positive variances. Otherwise, they will cancel each other out. That'd be bad.
    - Second, I also want to give more weight to the outliers, so this amplifies the effect of things that are very different from the mean while still, making sure that the negatives and positives are comparable *(11.56, 0.16, 0.36, 0.16, 12.96)*.

[67]

Let's look at what happens there, so $(-3.4)^2$ is a positive 11.56 and $(-0.4)^2$ ends up being a much smaller number, that is 0.16, because that's much closer to the mean of 4.4. Also $(0.6)^2$ turned out to be close to the mean, only 0.36. But as we get up to the positive outlier, $(3.6)^2$ ends up being 12.96. That gives us: *(11.56, 0.16, 0.36, 0.16, 12.96)*.

To find the actual variance value, we just take the average of all those squared differences. So we add up all these squared variances, divide the sum by 5, that is number of values that we have, and we end up with a variance of 5.04.

$$\sigma^2 = (11.56 + 0.16 + 0.36 + 0.16 + 12.96) / 5 = 5.04$$

OK, that's all variance is.

Standard deviation

Now typically, we talk about standard deviation more than variance, and it turns out standard deviation is just the square root of the variance. It's just that simple.

So, if I had this variance of *5.04*, the standard deviation is *2.24*. So you see now why we said that the variance = $(\sigma)^2$. It's because σ itself represents the standard deviation. So, if I take the square root of $(\sigma)^2$, I get sigma. That ends up in this example to be 2.24.

$$\sigma^2 = 5.04$$

$$\sigma = \sqrt{5.04} = 2.24$$

Chapter 2

Identifying outliers with standard deviation

Here's a histogram of the actual data we were looking at in the preceding example for calculating variance.

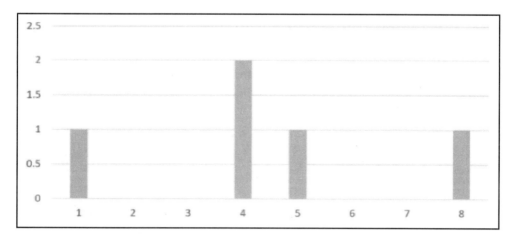

Now we see that the number **4** occurred twice in our dataset, and then we had one **1**, one **5**, and one **8**.

The standard deviation is usually used as a way to think about how to identify outliers in your dataset. If I say if I'm within one standard deviation of the mean of 4.4, that's considered to be kind of a typical value in a normal distribution. However, you can see in the preceding diagram, that the numbers **1** and **8** actually lie outside of that range. So if I take 4.4 plus or minus 2.24, we end up around **7** and **2**, and **1** and **8** both fall outside of that range of a standard deviation. So we can say mathematically, that 1 and 8 are outliers. We don't have to guess and eyeball it. Now there is still a judgment call as to what you consider an outlier in terms of how many standard deviations a data point is from the mean.

 You can generally talk about how much of an outlier a data point is by how many standard deviations (or sometimes how many-sigmas) from the mean it is.

So that's something you'll see standard deviation used for in the real world.

Population variance versus sample variance

There is a little nuance to standard deviation and variance, and that's when you're talking about population versus sample variance. If you're working with a complete set of data, a complete set of observations, then you do exactly what I told you. You just take the average of all the squared variances from the mean and that's your variance.

However, if you're sampling your data, that is, if you're taking a subset of the data just to make computing easier, you have to do something a little bit different. Instead of dividing by the number of samples, you divide by the number of samples minus 1. Let's look at an example.

We'll use the sample data we were just studying for people standing in a line. We took the sum of the squared variances and divided by 5, that is the number of data points that we had, to get 5.04.

$\sigma^2 = (11.56 + 0.16 + 0.36 + 0.16 + 12.96) / 5 = 5.04$

If we were to look at the sample variance, which is designated by S^2, it is found by the sum of the squared variances divided by 4, that is $(n - 1)$. This gives us the sample variance, which comes out to 6.3.

$S^2 = (11.56 + 0.16 + 0.36 + 0.16 + 12.96) / 4 = 6.3$

So again, if this was some sort of sample that we took from a larger dataset, that's what you would do. If it was a complete dataset, you divide by the actual number. Okay, that's how we calculate population and sample variance, but what's the actual logic behind it?

The Mathematical explanation

As for why there is difference between population and sample variance, it gets into really weird things about probability that you probably don't want to think about too much, and it requires some fancy mathematical notation, I try to avoid notation in this book as much as possible because I think the concepts are more important, but this is basic enough stuff and that you will see it over and over again.

As we've seen, population variance is usually designated as sigma squared (σ²), with sigma (σ) as standard deviation, and we can say that is the summation of each data point X minus the mean, mu, squared, that's the variance of each sample squared over N, the number of data points, and we can express it with the following equation:

$$\sigma^2 = \frac{\Sigma(X-\mu)^2}{N}$$

- X denotes each data point
- µ denotes the mean
- N denotes the number of data points

Sample variance similarly is designated as S², with the following equation:

$$S^2 = \frac{\Sigma(X-M)^2}{N-1}$$

- X denotes each data point
- M denotes the mean
- N-1 denotes the number of data points minus 1

That's all there is to it.

Analyzing standard deviation and variance on a histogram

Let's write some code here and play with some standard deviation and variances. So If you pull up the StdDevVariance.ipynb file IPython Notebook, and follow along with me here. Please do, because there's an activity at the end that I want you to try. What we're going to do here is just like the previous example, so begin with the following code:

```
%matplotlib inline
import numpy as np
import matplotlib.pyplot as plt
incomes = np.random.normal(100.0, 20.0, 10000)
plt.hist(incomes, 50)
plt.show()
```

We use matplotlib to plot a histogram of some normally distributed random data, and we call it incomes. We're saying it's going to be centered around 100 (hopefully that's an hourly rate or something and not annual, or some weird denomination), with a standard deviation of 20 and 10,000 data points.

Let's go ahead and generate that by executing that above code block and plotting it as shown in the following graph:

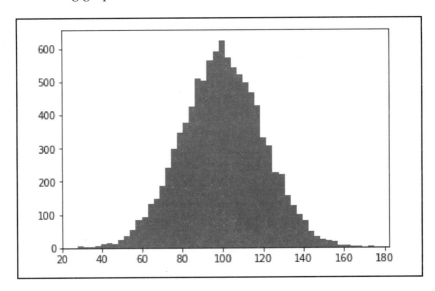

We have 10,000 data points centered around 100. With a normal distribution and a standard deviation of 20, a measure of the spread of this data, you can see that the most common occurrence is around 100, and as we get further and further from that, things become less and less likely. The standard deviation point of 20 that we specified is around 80 and around 120. You can see in the histogram that this is the point where things start to fall off sharply, so we can say that things beyond that standard deviation boundary are unusual.

Using Python to compute standard deviation and variance

Now, NumPy also makes it incredibly easy to compute the standard deviation and variance. If you want to compute the actual standard deviation of this dataset that we generated, you just call the `std` function right on the dataset itself. So, when NumPy creates the list, it's not just a normal Python list, it actually has some extra stuff tacked onto it so you can call functions on it, like `std` for standard deviation. Let's do that now:

```
incomes.std()
```

This gives us something like the following output (remember that we used random data, so your figures won't be exactly the same as mine):

```
20.024538249134373
```

When we execute that, we get a number pretty close to 20, because that's what we specified when we created our random data. We wanted a standard deviation of 20. Sure enough, 20.02, pretty close.

The variance is just a matter of calling `var`.

```
incomes.var()
```

This gives me the following:

```
400.98213209104557
```

It comes out to pretty close to 400, which is 20^2. Right, so the world makes sense! Standard deviation is just the square root of the variance, or you could say that the variance is the standard deviation squared. Sure enough, that works out, so the world works the way it should.

Try it yourself

I want you to dive in here and actually play around with it, make it real, so try out different parameters on generating that normal data. Remember, this is a measure of the shape of the distribution of the data, so what happens if I change that center point? Does it matter? Does it actually affect the shape? Why don't you try it out and find out?

Try messing with the actual standard deviation, that we've specified, to see what impact that has on the shape of the graph. Maybe try a standard deviation of 30, and you know, you can see how that actually affects things. Let's make it even more dramatic, like 50. Just play around with 50. You'll see the graph starting to get a little bit fatter. Play around with different values, just get a feel of how these values work. This is the only way to really get an intuitive sense of standard deviation and variance. Mess around with some different examples and see the effect that it has.

So that's standard deviation and variance in practice. You got hands on with some of it there, and I hope you played around a little bit to get some familiarity with it. These are very important concepts and we'll talk about standard deviations a lot throughout the book and no doubt throughout your career in data science, so make sure you've got that under your belt. Let's move on.

Probability density function and probability mass function

So we've already seen some examples of a normal distribution function for some of the examples in this book. That's an example of a probability density function, and there are other types of probability density functions out there. So let's dive in and see what it really means and what some other examples of them are.

The probability density function and probability mass functions

We've already seen some examples of a normal distribution function for some of the code we've looked at in this book. That's an example of a probability density function, and there are other types of probability density functions out there. Let's dive in and see what that really means and what some other examples of them there are.

Probability density functions

Let's talk about probability density functions, and we've used one of these already in the book. We just didn't call it that. Let's formalize some of the stuff that we've talked about. For example, we've seen the normal distribution a few times, and that is an example of a probability density function. The following figure is an example of a normal distribution curve

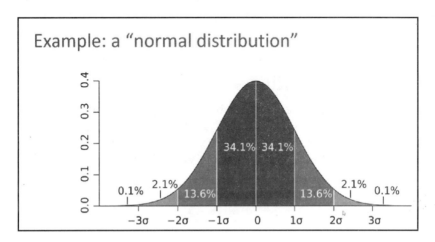

It's conceptually easy to try to think of this graph as the probability of a given value occurring, but that's a little bit misleading when you're talking about continuous data. Because there's an infinite number of actual possible data points in a continuous data distribution. There could be 0 or 0.001 or 0.00001 so the actual probability of a very specific value happening is very, very small, infinitely small. The probability density function really tells the probability of a given range of values occurring. So that's the way you've got to think about it.

So, for example, in the normal distribution shown in the above graph, between the mean (0) and one standard deviation from the mean (1σ) there's a **34.1%** chance of a value falling in that range. You can tighten this up or spread it out as much as you want, figure out the actual values, but that's the way to think about a probability density function. For a given range of values it gives you a way of finding out the probability of that range occurring.

- You can see in the graph, as you get close to the mean (0), within one standard deviation (**-1σ** and **1σ**), you're pretty likely to land there. I mean, if you add up 34.1 and 34.1, which equals to 68.2%, you get the probability of landing within one standard deviation of the mean.

[75]

- However, as you get between two and three standard deviations (**-3σ** to **-2σ** and **2σ** to **3σ**), we're down to just a little bit over 4% (4.2%, to be precise).
- As you get out beyond three standard deviations (**-3σ** and **3σ**) then we're much less than 1%.

So, the graph is just a way to visualize and talk about the probabilities of the given data point happening. Again, a probability distribution function gives you the probability of a data point falling within some given range of a given value, and a normal function is just one example of a probability density function. We'll look at some more in a moment.

Probability mass functions

Now when you're dealing with discrete data, that little nuance about having infinite numbers of possible values goes away, and we call that something different. So that is a probability mass function. If you're dealing with discrete data, you can talk about probability mass functions. Here's a graph to help visualize this:

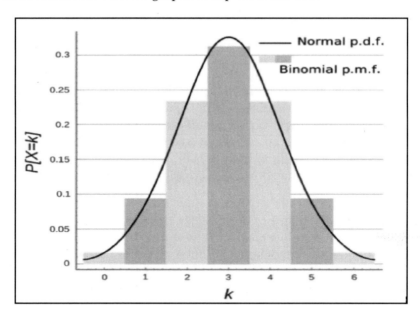

For example, you can plot a normal probability density function of continuous data on the black curve shown in the graph, but if we were to quantize that into a discrete dataset like we would do with a histogram, we can say the number 3 occurs some set number of times, and you can actually say the number 3 has a little over 30% chance of occurring. So a probability mass function is the way that we visualize the probability of discrete data occurring, and it looks a lot like a histogram because it basically is a histogram.

 Terminology difference: A probability density function is a solid curve that describes the probability of a range of values happening with continuous data. A probability mass function is the probabilities of given discrete values occurring in a dataset.

Types of data distributions

Let's look at some real examples of probability distribution functions and data distributions in general and wrap your head a little bit more around data distributions and how to visualize them and use them in Python.

Go ahead and open up the Distributions.ipynb from the book materials, and you can follow along with me here if you'd like.

Uniform distribution

Let's start off with a really simple example: uniform distribution. A uniform distribution just means there's a flat constant probability of a value occurring within a given range.

```
import numpy as np
Import matplotlib.pyplot as plt

values = np.random.uniform(-10.0, 10.0, 100000)
plt.hist(values, 50)
plt.show()
```

So we can create a uniform distribution by using the NumPy `random.uniform` function. The preceding code says, I want a uniformly distributed random set of values that ranges between `-10` and `10`, and I want `100000` of them. If I then create a histogram of those values, you can see it looks like the following.

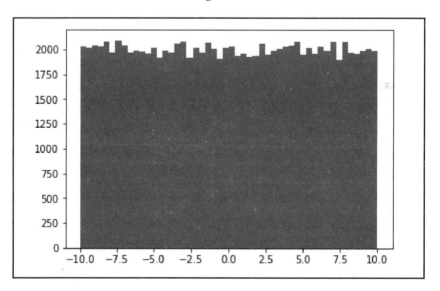

There's pretty much an equal chance of any given value or range of values occurring within that data. So, unlike the normal distribution, where we saw a concentration of values near the mean, a uniform distribution has equal probability across any given value within the range that you define.

So what would the probability distribution function of this look like? Well, I'd expect to see basically nothing outside of the range of **-10** or beyond **10**. But when I'm between **-10** and **10**, I would see a flat line because there's a constant probability of any one of those ranges of values occurring. So in a uniform distribution you would see a flat line on the probability distribution function because there is basically a constant probability. Every value, every range of values has an equal chance of appearing as any other value.

Normal or Gaussian distribution

Now we've seen normal, also known as Gaussian, distribution functions already in this book. You can actually visualize those in Python. There is a function called `pdf` (probability density function) in the `scipy.stats.norm` package function.

So, let's look at the following example:

```
from scipy.stats import norm
import matplotlib.pyplot as plt

x = np.arange(-3, 3, 0.001)
plt.plot(x, norm.pdf(x))
```

In the preceding example, we're creating a list of x values for plotting that range between -3 and 3 with an increment of 0.001 in between them by using the `arange` function. So those are the x values on the graph and we're going to plot the *x*-axis with using those values. The *y*-axis is going to be the normal function, `norm.pdf`, that the probability density function for a normal distribution, on those x values. We end up with the following output:

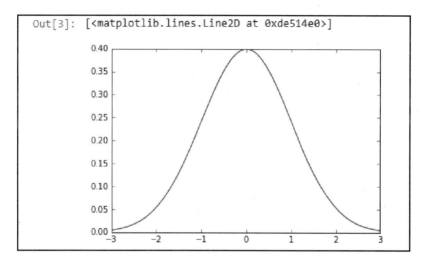

The pdf function with a normal distribution looks just like it did in our previous section, that is, a normal distribution for the given numbers that we provided, where 0 represents the mean, and the numbers **-3, -2, -1, 1, 2,** and **3** are standard deviations.

Now, we will generate random numbers with a normal distribution. We've done this a few times already; consider this a refresher. Refer to the following block of code:

```
import numpy as np
import matplotlib.pyplot as plt

mu = 5.0
sigma = 2.0
values = np.random.normal(mu, sigma, 10000)
plt.hist(values, 50)
plt.show()
```

In the above code, we use the `random.normal` function of the NumPy package, and the first parameter `mu`, represents the mean that you want to center the data around. `sigma` is the standard deviation of that data, which is basically the spread of it. Then, we specify the number of data points that we want using a normal probability distribution function, which is `10000` here. So that's a way to use a probability distribution function, in this case the normal distribution function, to generate a set of random data. We can then plot that, using a histogram broken into `50` buckets and show it. The following output is what we end up with:

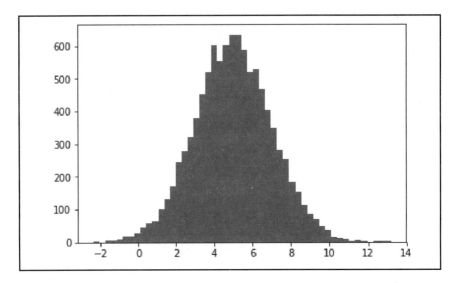

It does look more or less like a normal distribution, but since there is a random element, it's not going to be a perfect curve. We're talking about probabilities; there are some odds of things not quite being what they should be.

The exponential probability distribution or Power law

Another distribution function you see pretty often is the exponential probability distribution function, where things fall off in an exponential manner.

When you talk about an exponential fall off, you expect to see a curve, where it's very likely for something to happen, near zero, but then, as you get farther away from it, it drops off very quickly. There's a lot of things in nature that behave in this manner.

To do that in Python, just like we had a function in scipy.stats for norm.pdf, we also have an expon.pdf, or an exponential probability distribution function to do that in Python, we can do the same syntax that we did for the normal distribution with an exponential distribution here as shown in the following code block:

```
from scipy.stats import expon
import matplotlib.pyplot as plt

x = np.arange(0, 10, 0.001)
plt.plot(x, expon.pdf(x))
```

So again, in the above code, we just create our x values using the NumPy arange function to create a bunch of values between 0 and 10 with a step size of 0.001. Then, we plot those x values against the y-axis, which is defined as the function expon.pdf(x). The output looks like an exponential fall off. As shown in the following screenshot:

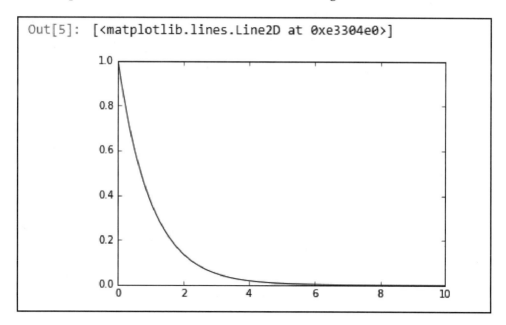

Binomial probability mass function

We can also visualize probability mass functions. This is called the binomial probability mass function. Again, we are going to use the same syntax as before, as shown in the following code:

```
from scipy.stats import expon
import matplotlib.pyplot as plt

x = np.arange(0, 10, 0.001)
plt.plot(x, expon.pdf(x))
```

So instead of `expon` or `norm`, we just use `binom`. A reminder: The probability mass function deals with discrete data. We have been all along, really, it's just how you think about it.

Coming back to our code, we're creating some discrete x values between 0 and 10 at a spacing of 0.01, and we're saying I want to plot a binomial probability mass function using that data. With the `binom.pmf` function, I can actually specify the shape of that data using two shape parameters, n and p. In this case, they're 10 and 0.5 respectively. output is shown on the following graph:

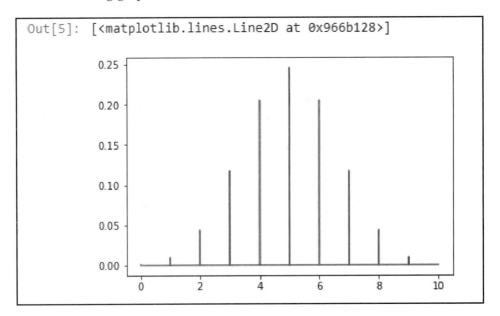

If you want to go and play around with different values to see what effects it has, that's a good way to get an intuitive sense of how those shape parameters work on the probability mass function.

Poisson probability mass function

Lastly, the other distribution function you might hear about is a Poisson probability mass function, and this has a very specific application. It looks a lot like a normal distribution, but it's a little bit different.

The idea here is, if you have some information about the average number of things that happen in a given time period, this probability mass function can give you a way to predict the odds of getting another value instead, on a given future day.

As an example, let's say I have a website, and on average I get 500 visitors per day. I can use the Poisson probability mass function to estimate the probability of seeing some other value on a specific day. For example, with my average of 500 visitors per day, what's the odds of seeing 550 visitors on a given day? That's what a Poisson probability mass function can give you take a look at the following code:

```
from scipy.stats import poisson
import matplotlib.pyplot as plt

mu = 500
x = np.arange(400, 600, 0.5)
plt.plot(x, poisson.pmf(x, mu))
```

In this code example, I'm saying my average is 500 mu. I'm going to set up some x values to look at between `400` and `600` with a spacing of `0.5`. I'm going to plot that using the `poisson.pmf` function. I can use that graph to look up the odds of getting any specific value that's not `500`, assuming a normal distribution:

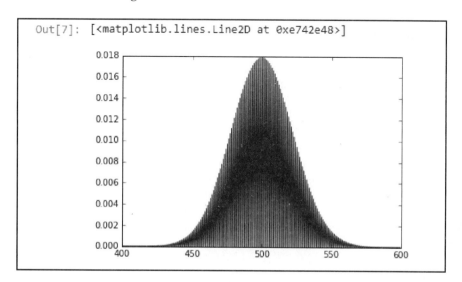

The odds of seeing **550** visitors on a given day, it turns out, comes out to about **0.002** or 0.2% probability. Very interesting.

Alright, so those are some common data distributions you might run into in the real world.

 Remember we used a probability distribution function with continuous data, but when we're dealing with discrete data, we use a probability mass function.

So that's probability density functions, and probability mass functions. Basically, a way to visualize and measure the actual chance of a given range of values occurring in a dataset. Very important information and a very important thing to understand. We're going to keep using that concept over and over again. Alright, let's move on.

Percentiles and moments

Next, we'll talk about percentiles and moments. You hear about percentiles in the news all the time. People that are in the top 1% of income: that's an example of percentile. We'll explain that and have some examples. Then, we'll talk about moments, a very fancy mathematical concept, but it turns out it's very simple to understand conceptually. Let's dive in and talk about percentiles and moments, a couple of a pretty basic concepts in statistics, but again, we're working our way up to the hard stuff, so bear with me as we go through some of this review.

Percentiles

Let's see what percentiles mean. Basically, if you were to sort all of the data in a dataset, a given percentile is the point at which that percent of the data is less than the point you're at.

A common example you see talked about a lot, is income distribution. When we talk about the 99th percentile, or the one-percenters, imagine that you were to take all the incomes of everybody in the country, in this case the United States, and sort them by income. The 99th percentile will be the income amount at which 99% of the rest of the country was making less than that amount. It's a very easy way to comprehend it.

 In a dataset, a percentile is the point at which $x\%$ of the values are less than the value at that point.

The following graph is an example for income distribution:

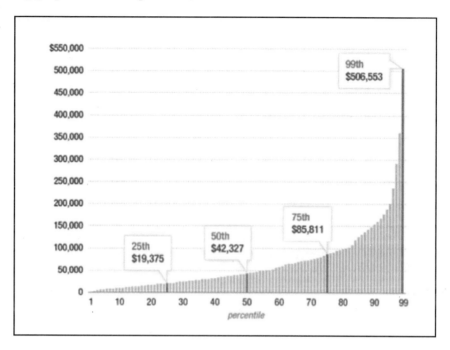

The preceding image shows an example of income distribution data. For example, at the 99th percentile we can say that 99% of the data points, which represent people in America, make less than $506,553 a year, and one percent make more than that. Conversely, if you're a one-percenter, you're making more than $506,553 a year. Congratulations! But if you're a more typical median person, the 50th percentile defines the point at which half of the people are making less and half are making more than you are, which is the definition of median. The 50th percentile is the same thing as median, and that would be at $42,327 given this dataset. So, if you're making $42,327 a year in the US, you are making exactly the median amount of income for the country.

You can see the problem of income distribution in the graph above. Things tend to be very concentrated toward the high end of the graph, which is a very big political problem right now in the country. We'll see what happens with that, but that's beyond the scope of this book. That's percentiles in a nutshell.

Quartiles

Percentiles are also used in the context of talking about the quartiles in a distribution. Let's look at a normal distribution to understand this better.

Here's an example illustrating Percentile in normal distribution:

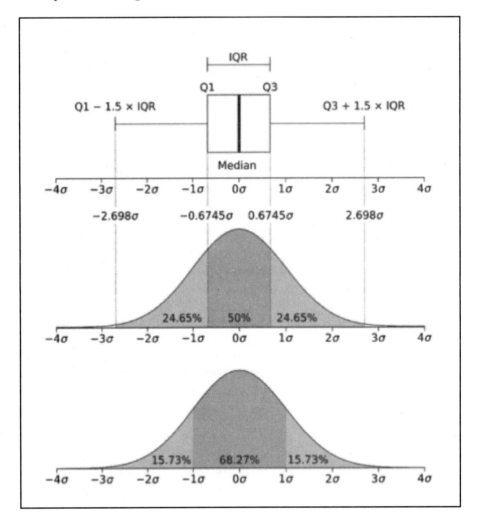

Looking at the normal distribution in the preceding image, we can talk about quartiles. Quartile 1 (Q1) and quartile 3 (Q3) in the middle are just the points that contain together 50% of the data, so 25% are on left side of the median and 25% are on the right side of the median.

The median in this example happens to be near the mean. For example, the **interquartile range (IQR)**, when we talk about a distribution, is the area in the middle of the distribution that contains 50% of the values.

The topmost part of the image is an example of what we call a box-and-whisker diagram. Don't concern yourself yet about the stuff out on the edges of the box. That gets a little bit confusing, and we'll cover that later. Even though they are called quartile 1 (Q1) and quartile 3 (Q1), they don't really represent 25% of the data, but don't get hung up on that yet. Focus on the point that the quartiles in the middle represent 25% of the data distribution.

Computing percentiles in Python

Let's look at some more examples of percentiles using Python and kind of get our hands on it and conceptualize this a little bit more. Go ahead and open the `Percentiles.ipynb` file if you'd like to follow along, and again I encourage you to do so because I want you to play around with this a little bit later.

Let's start off by generating some randomly distributed normal data, or normally distributed random data, rather, refer to the following code block:

```
%matplotlib inline
import numpy as np
import matplotlib.pyplot as plt

vals = np.random.normal(0, 0.5, 10000)

plt.hist(vals, 50)
plt.show()
```

In this example, what we're going to do is generate some data centered around zero, that is with a mean of zero, with a standard deviation of `0.5`, and I'm going to make `10000` data points with that distribution. Then, we're going to plot a histogram and see what we come up with.

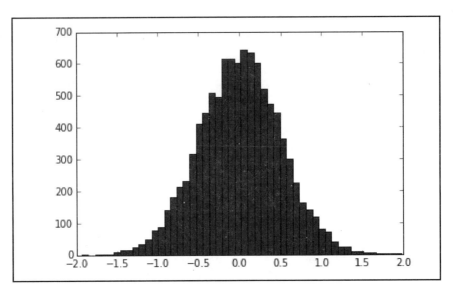

The generated histogram looks very much like a normal distribution, but because there is a random component we have a little outlier near the deviation of -2 in this example here. Things are tipped a little bit at the mean, a little bit of random variation there to make things interesting.

NumPy provides a very handy percentile function that will compute the percentile values of this distribution for you. So, we created our `vals` list of data using `np.random.normal`, and I can just call the `np.percentile` function to figure out the 50th percentile value in using the following code:

```
np.percentile(vals, 50)
```

The following is the output of the preceding code:

```
0.0053397035195310248
```

The output turns out to be 0.005. So remember, the 50th percentile is just another name for the median, and it turns out the median is very close to zero in this data. You can see in the graph that we're tipped a little bit to the right, so that's not too surprising.

I want to compute the 90th percentile, which gives me the point at which 90% of the data is less than that value. We can easily do that with the following code:

```
np.percentile(vals, 90)
```

Here is the output of that code:

```
Out[4]: 0.64099069837340827
```

The 90th percentile of this data turns out to be 0.64, so it's around here, and basically, at that point less than 90% of the data is less than that value. I can believe that. 10% of the data is greater than 0.64, 90% of it, less than 0.65.

Let's compute the 20th percentile value, that would give me the point at which 20% of the values are less than that number that I come up with. Again, we just need a very simple alteration to the code:

```
np.percentile(vals, 20)
```

This gives the following output:

```
Out[5]:-0.41810340026619164
```

The 20th percentile point works out to be -0.4, roughly, and again I believe that. It's saying that 20% of the data lies to the left of -0.4, and conversely, 80% is greater.

If you want to get a feel as to where those breaking points are in a dataset, the percentile function is an easy way to compute them. If this were a dataset representing income distribution, we could just call `np.percentile(vals, 99)` and figure out what the 99th percentile is. You could figure out who those one-percenters people keep talking about really are, and if you're one of them.

Alright, now to get your hands dirty. I want you to play around with this data. This is an IPython Notebook for a reason, so you can mess with it and mess with the code, try different standard deviation values, see what effect it has on the shape of the data and where those percentiles end up lying, for example. Try using smaller dataset sizes and add a little bit more random variation in the thing. Just get comfortable with it, play around with it, and find you can actually do this stuff and write some real code that works.

Moments

Next, let's talk about moments. Moments are a fancy mathematical phrase, but you don't actually need a math degree to understand it, though. Intuitively, it's a lot simpler than it sounds.

It's one of those examples where people in statistics and data science like to use big fancy terms to make themselves sound really smart, but the concepts are actually very easy to grasp, and that's the theme you're going to hear again and again in this book.

Basically, moments are ways to measure the shape of a data distribution, of a probability density function, or of anything, really. Mathematically, we've got some really fancy notation to define them:

$$\mu_n = \int_{-\infty}^{\infty} (x - c)^n f(x) dx \text{ (for moment } n \text{ around value } c\text{)}$$

If you do know calculus, it's actually not that complicated of a concept. We're taking the difference between each value from some value raised to the nth power, where n is the moment number and integrating across the entire function from negative infinity to infinity. But intuitively, it's a lot easier than calculus.

> Moments can be defined as quantitative measures of the shape of a probability density function.

Ready? Here we go!

1. The first moment works out to just be the mean of the data that you're looking at. That's it. The first moment is the mean, the average. It's that simple.
2. The second moment is the variance. That's it. The second moment of the dataset is the same thing as the variance value. It might seem a little bit creepy that these things kind of fall out of the math naturally, but think about it. The variance is really based on the square of the differences from the mean, so coming up with a mathematical way of saying that variance is related to mean isn't really that much of a stretch, right. It's just that simple.

3. Now when we get to the third and fourth moments, things get a little bit trickier, but they're still concepts that are easy to grasp. The third moment is called skew, and it is basically a measure of how lopsided a distribution is.

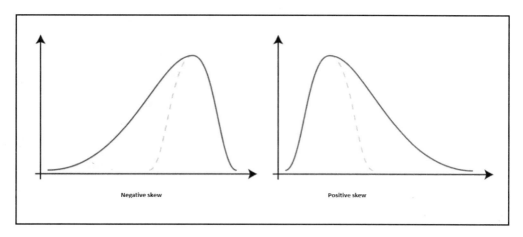

- You can see in these two examples above that, if I have a longer tail on the left, now then that is a negative skew, and if I have a longer tail on the right then, that's a positive skew. The dotted lines show what the shape of a normal distribution would look like without skew. The dotted line out on the left side then I end up with a negative skew, or on the other side, a positive skew in that example. OK, so that's all skew is. It's basically stretching out the tail on one side or the other, and it is a measure of how lopsided, or how skewed a distribution is.

4. The fourth moment is called kurtosis. Wow, that's a fancy word! All that really is, is how thick is the tail and how sharp is the peak. So again, it's a measure of the shape of the data distribution. Here's an example:

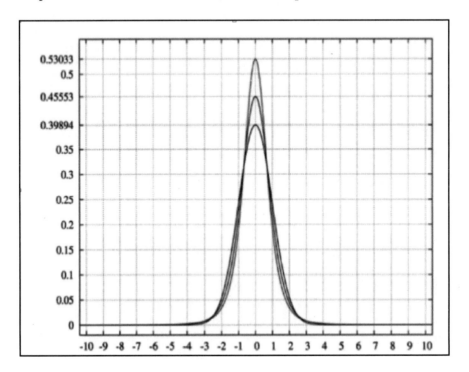

- You can see that the higher peak values have a higher kurtosis value. The topmost curve has a higher kurtosis than the bottommost curve. It's a very subtle difference, but a difference nonetheless. It basically measures how peaked your data is.

Let's review all that: the first moment is mean, the second moment is variance, the third moment is skew, and the fourth moment is kurtosis. We already know what mean and variance are. Skew is how lopsided the data is, how stretched out one of the tails might be. Kurtosis is how peaked, how squished together the data distribution is.

Computing moments in Python

Let's play around in Python and actually compute these moments and see how you do that. To play around with this, go ahead and open up the Moments.ipynb, and you can follow along with me here.

Let's again create that same normal distribution of random data. Again, we're going to make it centered around zero, with a 0.5 standard deviation and 10,000 data points, and plot that out:

```
import numpy as np
import matplotlib.pyplot as plt

vals = np.random.normal(0, 0.5, 10000)

plt.hist(vals, 50)
plt.show()
```

So again, we get a randomly generated set of data with a normal distribution around zero.

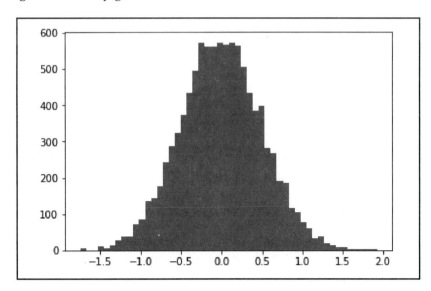

Now, we find the mean and variance. We've done this before; NumPy just gives you a mean and var function to compute that. So, we just call np.mean to find the first moment, which is just a fancy word for the mean, as shown in the following code:

```
np.mean(vals)
```

Statistics and Probability Refresher, and Python Practice

This gives the following output in our example:

```
Out [2]:-0.0012769999428169742
```

The output turns out to be very close to zero, just like we would expect for normally distributed data centered around zero. So, the world makes sense so far.

Now we find the second moment, which is just another name for variance. We can do that with the following code, as we've seen before:

```
np.var(vals)
```

Providing the following output:

```
Out[3]:0.25221246428323563
```

That output turns out to be about 0.25, and again, that works out with a nice sanity check. Remember that standard deviation is the square root of variance. If you take the square root of 0.25, it comes out to 0.5, which is the standard deviation we specified while creating this data, so again, that checks out too.

The third moment is skew, and to do that we're going to need to use the SciPy package instead of NumPy. But that again is built into any scientific computing package like Enthought Canopy or Anaconda. Once we have SciPy, the function call is as simple as our earlier two:

```
import scipy.stats as sp
sp.skew(vals)
```

This displays the following output:

```
Out[4]: 0.020055795996111746
```

We can just call `sp.skew` on the `vals` list, and that will give us the skew value. Since this is centered around zero, it should be almost a zero skew. It turns out that with random variation it does skew a little bit left, and actually that does jive with the shape that we're seeing in the graph. It looks like we did kind of pull it a little bit negative.

The fourth moment is kurtosis, which describes the shape of the tail. Again, for a normal distribution that should be about `zero`. SciPy provides us with another simple function call

```
sp.kurtosis(vals)
```

Chapter 2

And here's the output:

```
Out [5]:0.059954502386585506
```

Indeed, it does turn out to be zero. Kurtosis reveals our data distribution in two linked ways: the shape of the tail, or the how sharp the peak If I just squish the tail down it kind of pushes up that peak to be pointier, and likewise, if I were to push down that distribution, you can imagine that's kind of spreading things out a little bit, making the tails a little bit fatter, and the peak of it a little bit lower. So that's what kurtosis means, and in this example, kurtosis is near zero because it is just a plain old normal distribution.

If you want to play around with it, go ahead and, again, try to modify the distribution. Make it centered around something besides 0, and see if that actually changes anything. Should it? Well, it really shouldn't because these are all measures of the shape of the distribution, and it doesn't really say a whole lot about where that distribution is exactly. It's a measure of the shape. That's what the moments are all about. Go ahead and play around with that, try different center values, try different standard deviation values, and see what effect it has on these values, and it doesn't change at all. Of course, you'd expect things like the mean to change because you're changing the mean value, but variance, skew, maybe not. Play around, find out.

There you have percentiles and moments. Percentiles are a pretty simple concept. Moments sound hard, but it's actually pretty easy to understand how to do it, and it's easy in Python too. Now you have that under your belt. It's time to move on.

Summary

In this chapter, we saw the types of data (numeric, categorical, and ordinal data) that you might encounter and how to categorize them and how you treat them differently depending on what kind of data you're dealing with. We also walked through the statistical concepts of mean, median and mode, and we also saw the importance of choosing between median and mean, and that often the median is a better choice than the mean because of outliers.

Next, we analyzed how to compute mean, median, and mode using Python in an IPython Notebook file. We learned the concepts of standard deviation and variance in depth and how to compute them in Python. We saw that they're a measure of the spread of a data distribution. We also saw a way to visualize and measure the actual chance of a given range of values occurring in a dataset using probability density functions and probability mass functions.

We looked at the types of data distributions (Uniform distribution, Normal or Gaussian distribution, Exponential probability distribution, Binomial probability mass function, Poisson probability mass function) in general and how to visualize them using Python. We analyzed the concepts of percentiles and moments and saw how to compute them using Python.

In the next chapter, we'll look at using the `matplotlib` library more extensively, and also dive into the more advanced topics of covariance and correlation.

3
Matplotlib and Advanced Probability Concepts

After going through some of the simpler concepts of statistics and probability in the previous chapter, we're now going to turn our attention to some more advanced topics that you'll need to be familiar with to get the most out of the remainder of this book. Don't worry, they're not too complicated. First of all, let's have some fun and look at some of the amazing graphing capabilities of the `matplotlib` library.

We'll be covering the following topics in this chapter:

- Using the `matplotlib` package to plot graphs
- Understanding covariance and correlation to determine the relationship between data
- Understanding conditional probability with examples
- Understanding Bayes' theorem and its importance

A crash course in Matplotlib

Your data is only as good as you can present it to other people, really, so let's talk about plotting and graphing your data and how to present it to others and make your graphs look pretty. We're going to introduce Matplotlib more thoroughly and put it through its paces.

I'll show you a few tricks on how to make your graphs as pretty as you can. Let's have some fun with graphs. It's always good to make pretty pictures out of your work. This will give you some more tools in your tool chest for visualizing different types of data using different types of graphs and making it look pretty. We'll use different colors, different line styles, different axes, things like that. It's not only important to use graphs and data visualization to try to find interesting patterns in your data, but it's also interesting to present your findings well to a non-technical audience. Without further ado, let's dive in to Matplotlib.

Go ahead and open up the `MatPlotLib.ipynb` file and you can play around with this stuff along with me. We'll start by just drawing a simple line graph.

```
%matplotlib inline

from scipy.stats import norm
import matplotlib.pyplot as plt
import numpy as np

x = np.arange(-3, 3, 0.001)

plt.plot(x, norm.pdf(x))
plt.show()
```

So in this example, I import `matplotlib.pyplot` as `plt`, and with this, we can refer to it as `plt` from now on in this notebook. Then, I use `np.arange(-3, 3, 0.001)` to create an x-axis filled with values between -3 and 3 at increments of 0.001, and use `pyplot`'s `plot()` function to plot x. The y function will be `norm.pdf(x)`. So I'm going to create a probability density function with a normal distribution based on the x values, and I'm using the `scipy.stats norm` package to do that.

So tying it back into last chapter's look at probability density functions, here we are plotting a normal probability density function using `matplotlib`. So we just call `pyplot`'s `plot()` method to set up our plot, and then we display it using `plt.show()`. When we run the previous code, we get the following output:

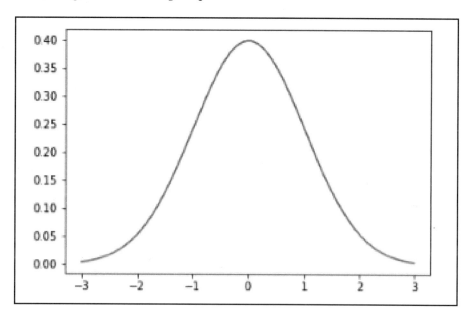

That's what we get: a pretty little graph with all the default formatting.

Generating multiple plots on one graph

Let's say I want to plot more than one thing at a time. You can actually call plot multiple times before calling show to actually add more than one function to your graph. Let's look at the following code:

```
plt.plot(x, norm.pdf(x))
plt.plot(x, norm.pdf(x, 1.0, 0.5))
plt.show()
```

Matplotlib and Advanced Probability Concepts

In this example, I'm calling my original function of just a normal distribution, but I'm going to render another normal distribution here as well, with a mean around 1.0 and a standard deviation of 0.5. Then, I'm going to show those two together so you can see how they compare to each other.

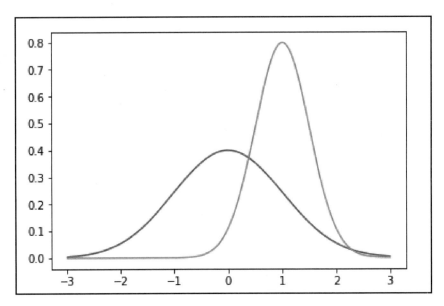

You can see that by default, matplotlib chooses different colors for each graph automatically for you, which is very nice and handy of it.

Saving graphs as images

If I want to save this graph to a file, maybe I want to include it in a document or something, I can do something like the following code:

```
plt.plot(x, norm.pdf(x))
plt.plot(x, norm.pdf(x, 1.0, 0.5))
plt.savefig('C:\\Users\\Frank\\MyPlot.png', format='png')
```

Instead of just calling plt.show(), I can call plt.savefig() with a path to where I want to save this file and what format I want it in.

You'll want to change that to an actual path that exists on your machine if you're following along. You probably don't have a `Users\Frank` folder on your system. Remember too that if you're on Linux or macOS, instead of a backslash you're going to use forward slashes, and you're not going to have a drive letter. With all of these Python Notebooks, whenever you see a path like this, make sure that you change it to an actual path that works on your system. I am on Windows here, and I do have a `Users\Frank` folder, so I can go ahead and run that. If I check my file system under `Users\Frank`, I have a `MyPlot.png` file that I can open up and look at, and I can use that in whatever document I want.

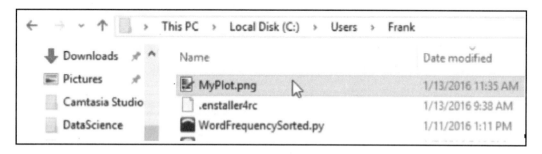

That's pretty cool. One other quick thing to note is that depending on your setup, you may have permissions issues when you come to save the file. You'll just need to find the folder that works for you. On Windows, your `Users\Name` folder is usually a safe bet. Alright, let's move on.

Adjusting the axes

Let's say that I don't like the default choices of the axes of this value in the previous graph. It's automatically fitting it to the tightest set of the axis values that you can find, which is usually a good thing to do, but sometimes you want things on an absolute scale. Look at the following code:

```
axes = plt.axes()
axes.set_xlim([-5, 5])
axes.set_ylim([0, 1.0])
axes.set_xticks([-5, -4, -3, -2, -1, 0, 1, 2, 3, 4, 5])
axes.set_yticks([0, 0.1, 0.2, 0.3, 0.4, 0.5, 0.6, 0.7, 0.8, 0.9, 1.0])
plt.plot(x, norm.pdf(x))
plt.plot(x, norm.pdf(x, 1.0, 0.5))
plt.show()
```

In this example, first I get the axes using `plt.axes`. Once I have these axes objects, I can adjust them. By calling `set_xlim`, I can set the x range from -5 to 5 and with set `set_ylim`, I set the y range from 0 to 1. You can see in the below output, that my x values are ranging from -5 to 5, and y goes from 0 to 1. I can also have explicit control over where the tick marks on the axes are. So in the previous code, I'm saying I want the x ticks to be at -5, -4, -3, etc., and y ticks from 0 to 1 at 0.1 increments using the `set_xticks()` and `set_yticks()` functions. Now I could use the `arange` function to do that more compactly, but the point is you have explicit control over where exactly those tick marks happen, and you can also skip some. You can have them at whatever increments you want or whatever distribution you want. Beyond that, it's the same thing.

Once I've adjusted my axes, I just called `plot()` with the functions that I want to plot and called `show()` to display it. Sure enough, there you have the result.

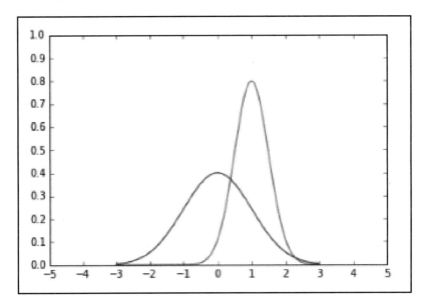

Adding a grid

What if I want grid lines in my graphs? Well, same idea. All I do is call `grid()` on the axes that I get back from `plt.axes()`.

```
axes = plt.axes()
axes.set_xlim([-5, 5])
axes.set_ylim([0, 1.0])
axes.set_xticks([-5, -4, -3, -2, -1, 0, 1, 2, 3, 4, 5])
```

```
axes.set_yticks([0, 0.1, 0.2, 0.3, 0.4, 0.5, 0.6, 0.7, 0.8, 0.9, 1.0])
axes.grid()
plt.plot(x, norm.pdf(x))
plt.plot(x, norm.pdf(x, 1.0, 0.5))
plt.show()
```

By executing the above code, I get nice little grid lines. That makes it a little bit easier to see where a specific point is, although it clutters things up a little bit. It's a little bit of a stylistic choice there.

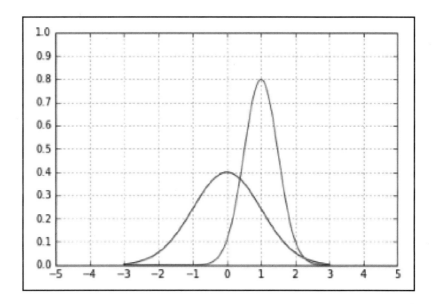

Changing line types and colors

What if I want to play games with the line types and colors? You can do that too.

```
axes = plt.axes()
axes.set_xlim([-5, 5])
axes.set_ylim([0, 1.0])
axes.set_xticks([-5, -4, -3, -2, -1, 0, 1, 2, 3, 4, 5])
axes.set_yticks([0, 0.1, 0.2, 0.3, 0.4, 0.5, 0.6, 0.7, 0.8, 0.9, 1.0])
axes.grid()
plt.plot(x, norm.pdf(x), 'b-')
plt.plot(x, norm.pdf(x, 1.0, 0.5), 'r:')
plt.show()
```

So you see in the preceding code, there's actually an extra parameter on the `plot()` functions at the end where I can pass a little string that describes the style of a line. In this first example, what `b-` indicates is I want a blue, solid line. The `b` stands for blue, and the dash means a solid line. For my second `plot()` function, I'm going to plot it in red, that's what the `r` means, and the colon means I'm going to plot it with a dotted line.

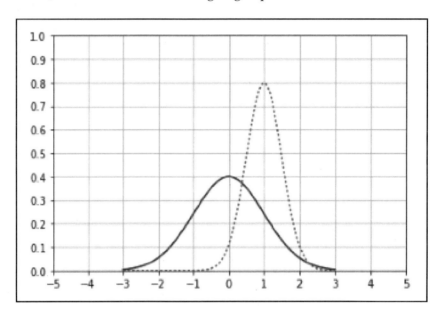

If I run that, you can see in the above graph what it does, and you can change different types of line styles.

In addition, you can do a double dash (`--`).

```
axes = plt.axes()
axes.set_xlim([-5, 5])
axes.set_ylim([0, 1.0])
axes.set_xticks([-5, -4, -3, -2, -1, 0, 1, 2, 3, 4, 5])
axes.set_yticks([0, 0.1, 0.2, 0.3, 0.4, 0.5, 0.6, 0.7, 0.8, 0.9, 1.0])
axes.grid()
plt.plot(x, norm.pdf(x), 'b-')
plt.plot(x, norm.pdf(x, 1.0, 0.5), 'r--')
plt.show()
```

The preceding code gives you dashed red line as a line style as shown in the following graph image:

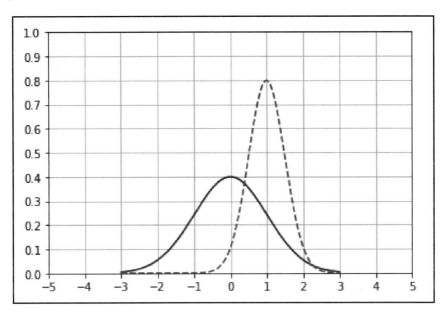

I can also do a dash dot combination (-.).

```
axes = plt.axes()
axes.set_xlim([-5, 5])
axes.set_ylim([0, 1.0])
axes.set_xticks([-5, -4, -3, -2, -1, 0, 1, 2, 3, 4, 5])
axes.set_yticks([0, 0.1, 0.2, 0.3, 0.4, 0.5, 0.6, 0.7, 0.8, 0.9, 1.0])
axes.grid()
plt.plot(x, norm.pdf(x), 'b-')
plt.plot(x, norm.pdf(x, 1.0, 0.5), 'r-.')
plt.show()
```

You get an output that looks like the following graph image:

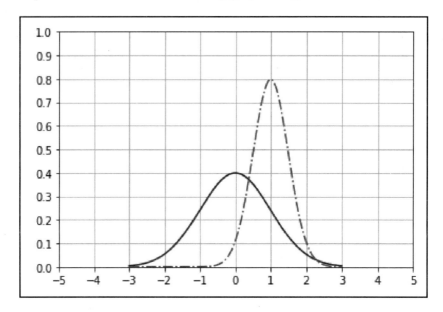

So, those are the different choices there. I could even make it green with vertical slashes (g:).

```
axes = plt.axes()
axes.set_xlim([-5, 5])
axes.set_ylim([0, 1.0])
axes.set_xticks([-5, -4, -3, -2, -1, 0, 1, 2, 3, 4, 5])
axes.set_yticks([0, 0.1, 0.2, 0.3, 0.4, 0.5, 0.6, 0.7, 0.8, 0.9, 1.0])
axes.grid()
plt.plot(x, norm.pdf(x), 'b-')
plt.plot(x, norm.pdf(x, 1.0, 0.5), ' g:')
plt.show()
```

I'll get the following output:

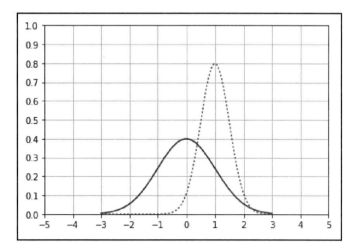

Have some fun with that if you want, experiment with different values, and you can get different line styles.

Labeling axes and adding a legend

Something you'll do more often is labeling your axes. You never want to present data in a vacuum. You definitely want to tell people what it represents. To do that, you can use the `xlabel()` and `ylabel()` functions on `plt` to actually put labels on your axes. I'll label the x axis Greebles and the y axis Probability. You can also add a legend inset. Normally, this would be the same thing, but just to show that it's set independently, I'm also setting up a legend in the following code:

```
axes = plt.axes()
axes.set_xlim([-5, 5])
axes.set_ylim([0, 1.0])
axes.set_xticks([-5, -4, -3, -2, -1, 0, 1, 2, 3, 4, 5])
axes.set_yticks([0, 0.1, 0.2, 0.3, 0.4, 0.5, 0.6, 0.7, 0.8, 0.9, 1.0])
axes.grid()
plt.xlabel('Greebles')
plt.ylabel('Probability')
plt.plot(x, norm.pdf(x), 'b-')
plt.plot(x, norm.pdf(x, 1.0, 0.5), 'r:')
plt.legend(['Sneetches', 'Gacks'], loc=4)
plt.show()
```

Matplotlib and Advanced Probability Concepts

Into the legend, you pass in basically a list of what you want to name each graph. So, my first graph is going to be called Sneetches, and my second graph is going to be called Gacks, and the `loc` parameter indicates what location you want it at, where 4 represents the lower right-hand corner. Let's go ahead and run the code, and you should see the following:

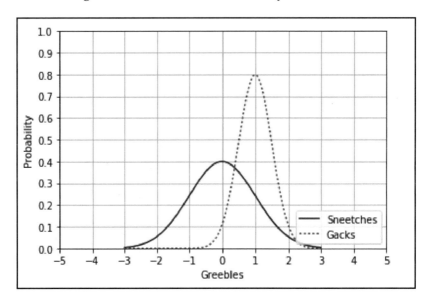

You can see that I'm plotting Greebles versus Probability for both Sneetches and Gacks. A little Dr. Seuss reference for you there. So that's how you set axes labels and legends.

A fun example

A little fun example here. If you're familiar with the webcomic XKCD, there's a little bit of an Easter egg in Matplotlib, where you can actually plot things in XKCD style. The following code shows how you can do that.

```
plt.xkcd()

fig = plt.figure()
ax = fig.add_subplot(1, 1, 1)
ax.spines['right'].set_color('none')
ax.spines['top'].set_color('none')
plt.xticks([])
plt.yticks([])
ax.set_ylim([-30, 10])

data = np.ones(100)
```

```
data[70:] -= np.arange(30)

plt.annotate(
    'THE DAY I REALIZED\nI COULD COOK BACON\nWHENEVER I WANTED',
    xy=(70, 1), arrowprops=dict(arrowstyle='->'), xytext=(15, -10))

plt.plot(data)

plt.xlabel('time')
plt.ylabel('my overall health')
```

In this example, you call plt.xkcd(), which puts Matplotlib in XKCD mode. After you do that, things will just have a style with kind of a comic book font and squiggly lines automatically. This little simple example will show a funny little graph where we are plotting your health versus time, where your health takes a steep decline once you realize you can cook bacon whenever you want to. All we're doing there is using the xkcd() method to go into that mode. You can see the results below:

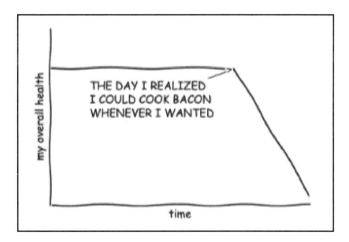

There's a little bit of interesting Python here in how we're actually putting this graph together. We're starting out by making a data line that is nothing but the value 1 across 100 data points. Then we use the old Python list slicing operator to take everything after the value of 70, and we subtract off from that sub-list of 30 items, the range of 0 through 30. So that has the effect of subtracting off a larger value linearly as you get past 70, which results in that line heading downward down to 0 beyond the point 70.

So, it's a little example of some Python list slicing in action there, and a little creative use of the arange function to modify your data.

Generating pie charts

Now, to go back to the real world, we can remove XKCD mode by calling `rcdefaults()` on Matplotlib, and we can get back to normal mode here.

If you want a pie chart, all you have to do is call `plt.pie` and give it an array of your values, colors, labels, and whether or not you want items exploded, and if so, by how much. Here's the code:

```
# Remove XKCD mode:
plt.rcdefaults()

values = [12, 55, 4, 32, 14]
colors = ['r', 'g', 'b', 'c', 'm']
explode = [0, 0, 0.2, 0, 0]
labels = ['India', 'United States', 'Russia', 'China', 'Europe']
plt.pie(values, colors= colors, labels=labels, explode = explode)
plt.title('Student Locations')
plt.show()
```

You can see in this code that I'm creating a pie chart with the values 12, 55, 4, 32, and 14. I'm assigning explicit colors to each one of those values, and explicit labels to each one of those values. I'm exploding out the Russian segment of the pie by 20%, and giving this plot a title of Student Locations and showing it. The following is the output you should see:

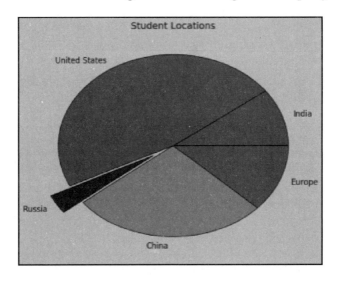

That's all there is to it.

Generating bar charts

If I want to generate a bar chart, that is also very simple. It's a kind of a similar idea to the pie chart. Let's look at the following code.

```
values = [12, 55, 4, 32, 14]
colors = ['r', 'g', 'b', 'c', 'm']
plt.bar(range(0,5), values, color= colors)
plt.show()
```

I've defined an array of values and an array of colors, and just plot the data. The above code plots from the range of 0 to 5, using the y values from the `values` array and using the explicit list of colors listed in the `colors` array. Go ahead and show that, and there you have your bar chart:

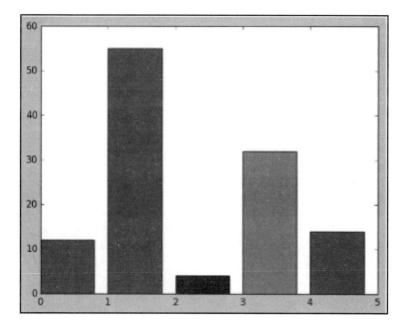

Generating scatter plots

A scatter plot is something we'll see pretty often in this book. So, say you have a couple of different attributes you want to plot for the same set of people or things. For example, maybe we're plotting ages against incomes for each person, where each dot represents a person and the axes represent different attributes of those people.

The way you do that with a scatter plot is you call plt.scatter() using the two axes that you want to define, that is, the two attributes that contain data that you want to plot against each other.

Let's say I have a random distribution in X and Y and I scatter those on the scatter plot, and I show it:

```
from pylab import randn

X = randn(500)
Y = randn(500)
plt.scatter(X,Y)
plt.show()
```

You get the following scatter plot as output:

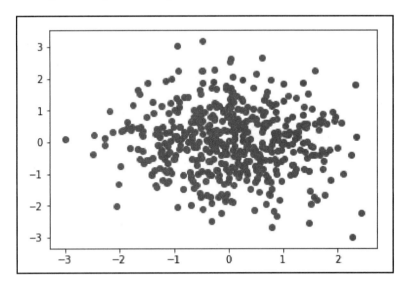

This is what it looks like, pretty cool. You can see the sort of a concentration in the center here, because of the normal distribution that's being used in both axes, but since it is random, there's no real correlation between those two.

Generating histograms

Finally, we'll remind ourselves how a histogram works. We've already seen this plenty of times in the book. Let's look at the following code:

```
incomes = np.random.normal(27000, 15000, 10000)
plt.hist(incomes, 50)
plt.show()
```

In this example, I call a normal distribution centered on 27,000, with a standard deviation of 15,000 with 10,000 data points. Then, I just call `pyplot`'s histogram function, that is, `hist()`, and specify the input data and the number of buckets that we want to group things into in our histogram. Then I call `show()` and the rest is magic.

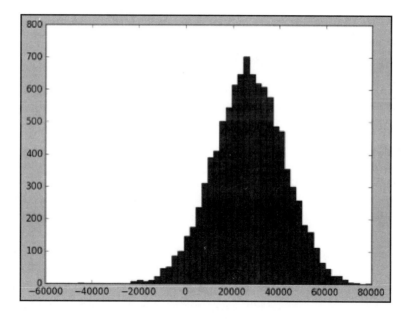

Generating box-and-whisker plots

Finally, let's look at box-and-whisker plots. Remember in the previous chapter, when we talked about percentiles I touched on this a little bit.

Again, with a box-and-whisker plot, the box represents the two inner quartiles where 50% of your data resides. Conversely, another 25% resides on either side of that box; the whiskers (dotted lines in our example) represent the range of the data except for outliers.

Matplotlib and Advanced Probability Concepts

We define outliers in a box-and-whisker plot as anything beyond 1.5 times the interquartile range, or the size of the box. So, we take the size of the box times 1.5, and up to that point on the dotted whiskers, we call those parts outer quartiles. But anything outside of the outer quartiles is considered an outlier, and that's what the lines beyond the outer quartiles represent. That's where we are defining outliers based on our definition with the box-and-whisker plot.

Some points to remember about box-and-whisker plots:

- They are useful for visualizing the spread and skew of data
- The line in the middle of the box represents the median of the data, and the box represents the bounds of the 1st and 3rd quartiles
- Half of the data exists within the box
- The "whiskers" indicate the range of the data-except for outliers, which are plotted outside the whiskers.
- Outliers are 1.5 times or more the interquartile range.

Now, just to give you an example here, we have created a fake dataset. The following example creates uniformly distributed random numbers between -40 and 60, plus a few outliers above `100` and below `-100`:

```
uniformSkewed = np.random.rand(100) * 100 - 40
high_outliers = np.random.rand(10) * 50 + 100
low_outliers = np.random.rand(10) * -50 - 100
data = np.concatenate((uniformSkewed, high_outliers, low_outliers))
plt.boxplot(data)
plt.show()
```

In the code, we have a uniform random distribution of data (`uniformSkewed`). Then we added a few outliers on the high end (`high_outliers`) and a few negative outliers (`low_outliers`) as well. Then we concatenated these lists together and created a single dataset from these three different sets that we created using NumPy. We then took that combined dataset of uniform data and a few outliers and we plotted using `plt.boxplot()`, and that's how you get a box-and-whisker plot. Call `show()` to visualize it, and there you go.

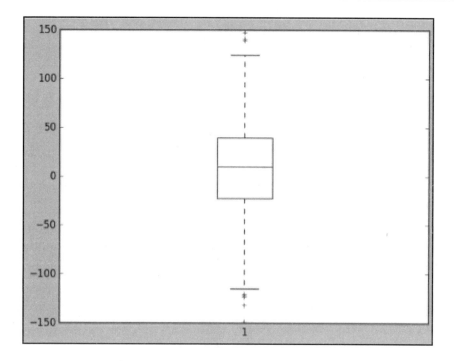

You can see that the graph is showing the box that represents the inner 50% of all data, and then we have these outlier lines where we can see little crosses (they may be circles in your version) for each individual outlier that lies in that range.

Try it yourself

Alright, that's your crash course in Matplotlib. Time to get your hands on it, and actually do some exercises here.

As your challenge, I want you to create a scatter plot that represents random data that you fabricate on age versus time spent watching TV, and you can make anything you want, really. If you have a different fictional data set in your head that you like to play with, have some fun with it. Create a scatter plot that plots two random sets of data against each other and label your axes. Make it look pretty, play around with it, have fun with it. Everything you need for reference and for examples should be in this IPython Notebook. It's kind of a cheat sheet, if you will, for different things you might need to do for generating different kinds of graphs and different styles of graphs. I hope it proves useful. Now it's time to get back to the statistics.

Covariance and correlation

Next, we're going to talk about covariance and correlation. Let's say I have two different attributes of something and I want to see if they're actually related to each other or not. This section will give you the mathematical tools you need to do so, and we'll dive into some examples and actually figure out covariance and correlation using Python. These are ways of measuring whether two different attributes are related to each other in a set of data, which can be a very useful thing to find out.

Defining the concepts

Imagine we have a scatter plot, and each one of the data points represents a person that we measured, and we're plotting their age on one axis versus their income on another. Each one of these dots would represent a person, for example their x value represents their age and the y value represents their income. I'm totally making this up, this is fake data.

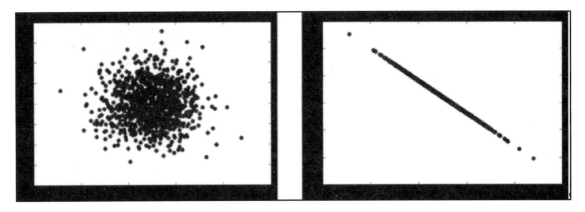

Now if I had a scatter plot that looks like the left one in the preceding image, you see that these values tend to lie all over the place, and this would tell you that there's no real correlation between age and income based on this data. For any given age, there can be a huge range of incomes and they tend to be clustered around the middle, but we're not really seeing a very clear relationship between these two different attributes of age and income. Now in contrast, in the scatter plot on the right you can see there's a very clear linear relationship between age and income.

So, covariance and correlation give us a means of measuring just how tight these things are correlated. I would expect a very low correlation or covariance for the data in the left scatter plot, but a very high covariance and correlation for the data in the right scatter plot. So that's the concept of covariance and correlation. It measures how much these two attributes that I'm measuring seem to depend on each other.

Measuring covariance

Measuring covariance mathematically is a little bit hard, but I'll try to explain it. These are the steps:

- Think of the data sets for the two variables as high-dimensional vectors
- Convert these to vectors of variances from the mean
- Take the dot product (cosine of the angle between them) of the two vectors
- Divide by the sample size

It's really more important that you understand how to use it and what it means. To actually derive it, think of the attributes of the data as high dimensional vectors. What we're going to do on each attribute for each data point is compute the variance from the mean at each point. So now I have these high dimensional vectors where each data point, each person, if you will, corresponds to a different dimension.

I have one vector in this high dimensional space that represents all the variances from the mean for, let's say, age for one attribute. Then I have another vector that represents all the variances from the mean for some other attribute, like income. What I do then is I take these vectors that measure the variances from the mean for each attribute, and I take the dot product between the two. Mathematically, that's a way of measuring the angle between these high dimensional vectors. So if they end up being very close to each other, that tells me that these variances are pretty much moving in lockstep with each other across these different attributes. If I take that final dot product and divide it by the sample size, that's how I end up with the covariance amount.

Now you're never going to have to actually compute this yourself the hard way. We'll see how to do this the easy way in Python, but conceptually, that's how it works.

Now the problem with covariance is that it can be hard to interpret. If I have a covariance that's close to zero, well, I know that's telling me there's not much correlation between these variables at all, but a large covariance implies there is a relationship. But how large is large? Depending on the units I'm using, there might be very different ways of interpreting that data. That's a problem that correlation solves.

Correlation

Correlation normalizes everything by the standard deviation of each attribute (just divide the covariance by the standard deviations of both variables and that normalizes things). By doing so, I can say very clearly that a correlation of -1 means there's a perfect inverse correlation, so as one value increases, the other decreases, and vice versa. A correlation of 0 means there's no correlation at all between these two sets of attributes. A correlation of 1 would imply perfect correlation, where these two attributes are moving in exactly the same way as you look at different data points.

Remember, correlation does not imply causation. Just because you find a very high correlation value does not mean that one of these attributes causes the other. It just means there's a relationship between the two, and that relationship could be caused by something completely different. The only way to really determine causation is through a controlled experiment, which we'll talk about more later.

Computing covariance and correlation in Python

Alright, let's get our hands dirty with covariance and correlation here with some actual Python code. So again, you can think conceptually of covariance as taking these multi-dimensional vectors of variances from the mean for each attribute and computing the angle between them as a measure of the covariance. The math for doing that is a lot simpler than it sounds. We're talking about high dimensional vectors. It sounds like Stephen Hawking stuff, but really, from a mathematical standpoint it's pretty straightforward.

Computing correlation – The hard way

I'm going to start by doing this the hard way. NumPy does have a method to just compute the covariance for you, and we'll talk about that later, but for now I want to show that you can actually do this from first principles:

```
%matplotlib inline

import numpy as np
from pylab import *

def de_mean(x):
    xmean = mean(x)
    return [xi - xmean for xi in x]

def covariance(x, y):
```

```
n = len(x)
return dot(de_mean(x), de_mean(y)) / (n-1)
```

Covariance, again, is defined as the dot product, which is a measure of the angle between two vectors, of a vector of the deviations from the mean for a given set of data and the deviations from the mean for another given set of data for the same data's data points. We then divide that by n - 1 in this case, because we're actually dealing with a sample.

So `de_mean()`, our deviation from the mean function is taking in a set of data, x, actually a list, and it's computing the mean of that set of data. The `return` line contains a little bit of Python trickery for you. The syntax is saying, I'm going to create a new list, and go through every element in x, call it `xi`, and then return the difference between `xi` and the mean, `xmean`, for that entire dataset. This function returns a new list of data that represents the deviations from the mean for each data point.

My `covariance()` function will do that for both sets of data coming in, divided by the number of data points minus 1. Remember that thing about sample versus population in the previous chapter? Well, that's coming into play here. Then we can just use those functions and see what happens.

To expand this example, I'm going to fabricate some data that is going to try to find a relationship between page speeds, that, is how quickly a page renders on a website, and how much people spend. For example, at Amazon we were very concerned about the relationship between how quickly pages render and how much money people spend after that experience. We wanted to know if there is an actual relationship between how fast the website is and how much money people actually spend on the website. This is one way you might go about figuring that out. Let's just generate some normally distributed random data for both page speeds and purchase amounts, and since it's random, there's not going to be a real correlation between them.

```
pageSpeeds = np.random.normal(3.0, 1.0, 1000)
purchaseAmount = np.random.normal(50.0, 10.0, 1000)

scatter(pageSpeeds, purchaseAmount)

covariance (pageSpeeds, purchaseAmount)
```

Matplotlib and Advanced Probability Concepts

So just as a sanity check here we'll start off by scatter plotting this stuff:

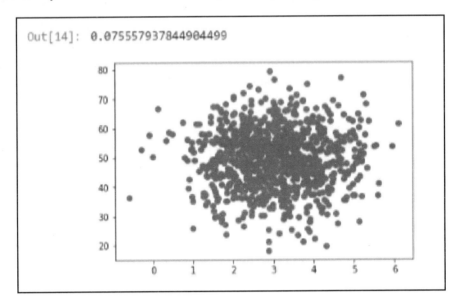

You'll see that it tends to cluster around the middle because of the normal distribution on each attribute, but there's no real relationship between the two. For any given page speed is a wide variety of amount spent, and for any given amount spent there's a wide variety of page speeds, so no real correlation there except for ones that are coming out the randomness or through the nature of the normal distribution. Sure enough, if we compute the covariance in these two sets of attributes, we end up with a very small value, -0.07. So that's a very small covariance value, close to zero. That implies there's no real relationship between these two things.

Now let's make life a little bit more interesting. Let's actually make the purchase amount a real function of page speed.

```
purchaseAmount = np.random.normal(50.0, 10.0, 1000) / pageSpeeds

scatter(pageSpeeds, purchaseAmount)

covariance (pageSpeeds, purchaseAmount)
```

Chapter 3

Here, we are keeping things a little bit random, but we are creating a real relationship between these two sets of values. For a given user, there's a real relationship between the page speeds they encounter and the amount that they spend. If we plot that out, we can see the following output:

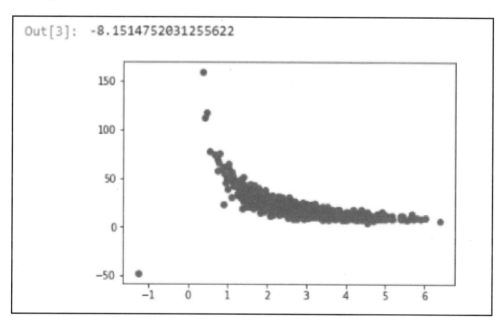

You can see that there's actually this little curve where things tend to be tightly aligned. Things get a little bit wonky near the bottom, just because of how random things work out. If we compute the covariance, we end up with a much larger value, -8, and it's the magnitude of that number that matters. The sign, positive or negative, just implies a positive or negative correlation, but that value of 8 says that's a much higher value than zero. So there's something going on there, but again it's hard to interpret what 8 actually means.

That's where the correlation comes in, where we normalize everything by the standard deviations as shown in the following code:

```
def correlation(x, y):
stddevx = x.std()
stddevy = y.std()
return covariance(x,y) / stddevx / stddevy   #In real life you'd check for divide by zero here

correlation(pageSpeeds, purchaseAmount)
```

Again, doing that from first principles, we can take the correlation between two sets of attributes, compute the standard deviation of each, then compute the covariance between these two things, and divide by the standard deviations of each dataset. That gives us the correlation value, which is normalized to -1 to 1. We end up with a value of -0.4, which tells us there is some correlation between these two things in the negative direction:

```
Out[3]:   -0.46775563114087165
```

It's not a perfect line, that would be -1, but there's something interesting going on there.

 A -1 correlation coefficient means perfect negative correlation, 0 means no correlation, and 1 means perfect positive correlation.

Computing correlation – The NumPy way

Now, NumPy can actually compute correlation for you using the `corrcoef()` function. Let's look at the following code:

```
np.corrcoef(pageseeds, purchaseAmount)
```

This single line gives the following output:

```
array([[ 1.         ,-046728788],
       [-0.46728788], 1.         ])
```

So, if we wanted to do this the easy way, we could just use `np.corrcoef(pageSpeeds, purchaseAmount)`, and what that gives you back is an array that gives you the correlation between every possible combination of the sets of data that you pass in. The way to read the output is: the 1 implies there is a perfect correlation between comparing `pageSpeeds` to itself and `purchaseAmount` to itself, which is expected. But when you start comparing `pageSpeeds` to `purchaseAmount` or `purchaseAmount` to the `pageSpeeds`, you end up with the -0.4672 value, which is roughly what we got when we did it the hard way. There's going to be little precision errors, but it's not really important.

Now we could force a perfect correlation by fabricating a totally linear relationship, so let's take a look at an example of that:

```
purchaseAmount = 100 - pageSpeeds * 3

scatter(pageSpeeds, purchaseAmount)

correlation (pageSpeeds, purchaseAmount)
```

And again, here we would expect the correlation to come out to -1 for a perfect negative correlation, and in fact, that's what we end up with:

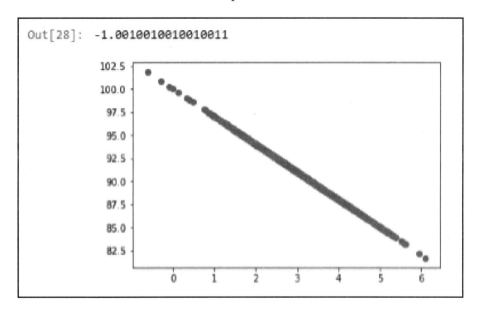

Again, a reminder: Correlation does not imply causality. Just because people might spend more if they have faster page speeds, maybe that just means that they can afford a better Internet connection. Maybe that doesn't mean that there's actually a causation between how fast your pages render and how much people spend, but it tells you there's an interesting relationship that's worth investigating more. You cannot say anything about causality without running an experiment, but correlation can tell you what experiments you might want to run.

Correlation activity

So get your hands dirty, roll up your sleeves, I want you to use the `numpy.cov()` function. That's actually a way to get NumPy to compute covariance for you. We saw how to compute correlation using the `corrcoef()` function. So go back and rerun these examples just using the `numpy.cov()` function and see if you get the same results or not. It should be pretty darn close, so instead of doing it the hard way with the covariance function that I wrote from scratch, just use NumPy and see if you can get the same results. Again, the point of this exercise is to get you familiar with using NumPy and applying it to actual data. So have at it, see where you get.

And there you have it, covariance and correlation both in theory and in practice. A very useful technique to have, so definitely remember this section. Let's move on.

Conditional probability

Next, we're going to talk about conditional probability. It's a very simple concept. It's trying to figure out the probability of something happening given that something else occurred. Although it sounds simple, it can be actually very difficult to wrap your head around some of the nuances of it. So get an extra cup of coffee, make sure your thinking cap's on, and if you're ready for some more challenging concepts here. Let's do this.

Conditional probability is a way to measure the relationship between two things happening to each other. Let's say I want to find the probability of an event happening given that another event already happened. Conditional probability gives you the tools to figure that out.

What I'm trying to find out with conditional probability is if I have two events that depend on each other. That is, what's the probability that both will occur?

In mathematical notation, the way we indicate things here is that *P(A,B)* represents the probability of both A and B occurring independent of each other. That is, what's the probability of both of these things happening irrespective of everything else.

Whereas this notation, *P(B|A)*, is read as the probability of B given A. So, what is the probability of B given that event A has already occurred? It's a little bit different, and these things are related like this:

$$P(B|A) = \frac{P(A,B)}{P(A)}$$

The probability of B given A is equal to the probability of A and B occurring over the probability of A alone occurring, so this teases out the probability of B being dependent on the probability of A.

It'll make more sense with an example here, so bear with me.

Let's say that I give you, my readers, two tests, and 60% of you pass both tests. Now the first test was easier, 80% of you passed that one. I can use this information to figure out what percentage of readers who passed the first test also passed the second. So here's a real example of the difference between the probability of B given A and the probability of A and B.

I'm going to represent A as the probability of passing the first test, and B as the probability of passing the second test. What I'm looking for is the probability of passing the second test given that you passed the first, that is, $P(B|A)$.

$$P(B|A) = \frac{P(A,B)}{P(A)} = \frac{0.6}{0.8} = 0.75$$

So the probability of passing the second test given that you passed the first is equal to the probability of passing both tests, $P(A,B)$ (I know that 60% of you passed both tests irrespective of each other), divided by the probability of passing the first test, $P(A)$, which is 80%. It's worked out to 60% passed both tests, 80% passed the first test, therefore the probability of passing the second given that you passed the first works out to 75%.

OK, it's a little bit tough to wrap your head around this concept. It took me a little while to really internalize the difference between the probability of something given something and the probability of two things happening irrespective of each other. Make sure you internalize this example and how it's really working before you move on.

Conditional probability exercises in Python

Alright, let's move on and do another more complicated example using some real Python code. We can then see how we might actually implement these ideas using Python.

Let's put conditional probability into action here and use some of the ideas to figure out if there's a relationship between age and buying stuff using some fabricated data. Go ahead and open up the `ConditionalProbabilityExercise.ipynb` here and follow along with me if you like.

What I'm going to do is write a little bit of Python code that creates some fake data:

```
from numpy import random
random.seed(0)

totals = {20:0, 30:0, 40:0, 50:0, 60:0, 70:0}
purchases = {20:0, 30:0, 40:0, 50:0, 60:0, 70:0}
totalPurchases = 0
for _ in range(100000):
    ageDecade = random.choice([20, 30, 40, 50, 60, 70])
    purchaseProbability = float(ageDecade) / 100.0
    totals[ageDecade] += 1
    if (random.random() < purchaseProbability):
        totalPurchases += 1
        purchases[ageDecade] += 1
```

What I'm going to do is take 100,000 virtual people and randomly assign them to an age bracket. They can be in their 20s, their 30s, their 40s, their 50s, their 60s, or their 70s. I'm also going to assign them a number of things that they bought during some period of time, and I'm going to weight the probability of purchasing something based on their age.

What this code ends up doing is randomly assigning each person to an age group using the `random.choice()` function from NumPy. Then I'm going to assign a probability of purchasing something, and I have weighted it such that younger people are less likely to buy stuff than older people. I'm going to go through 100,000 people and add everything up as I go, and what I end up with are two Python dictionaries: one that gives me the total number of people in each age group, and another that gives me the total number of things bought within each age group. I'm also going to keep track of the total number of things bought overall. Let's go ahead and run that code.

If you want to take a second to kind of work through that code in your head and figure out how it works, you've got the IPython Notebook. You can go back into that later too. Let's take a look what we ended up with.

```
In [2]: totals
Out[2]: {20: 16576, 30: 16619, 40: 16632, 50: 16805, 60: 16664, 70: 16704}

In [3]: purchases
Out[3]: {20: 3392, 30: 4974, 40: 6670, 50: 8319, 60: 9944, 70: 11713}

In [4]: totalPurchases
Out[4]: 45012
```

Our `totals` dictionary is telling us how many people are in each age bracket, and it's pretty evenly distributed, just like we expected. The amount purchased by each age group is in fact increasing by age, so 20-year-olds only bought about 3,000 things and 70-year-olds bought about 11,000 things, and overall the entire population bought about 45,000 things.

Let's use this data to play around with the ideas of conditional probability. Let's first figure out what's the probability of buying something given that you're in your 30s. The notation for that will be $P(E|F)$ if we're calling purchase E, and F as the event that you're in your 30s.

Now we have this fancy equation that gave you a way of computing $P(E|F)$ given $P(E,F)$, and $P(E)$, but we don't need that. You don't just blindly apply equations whenever you see something. You have to think about your data intuitively. What is it telling us? I want to figure out the probability of purchasing something given that you're in your 30s. Well I have all the data I need to compute that directly.

```
PEF = float(purchases[30]) / float(totals[30])
```

I have how much stuff 30-year-olds purchased in the purchases[30] bucket, and I know how many 30-year-olds there are. So I can just divide those two numbers to get the ratio of 30-year-old purchases over the number of 30-year-olds. I can then output that using the print command:

```
print ("P(purchase | 30s): ", PEF)
```

I end up with a probability of purchasing something given that you're in your 30s of being about 30%:

```
P(purchase | 30s): 0.2992959865211
```

Note that if you're using Python 2, the print command doesn't have the surrounding brackets, so it would be:

```
print "p(purchase | 30s): ", PEF
```

If I want to find $P(F)$, that's just the probability of being 30 overall, I can take the total number of 30-year-olds divided by the number of people in my dataset, which is 100,000:

```
PF = float(totals[30]) / 100000.0
print ("P(30's): ", PF)
```

Again, remove those brackets around the print statement if you're using Python 2. That should give the following output:

```
P(30's): 0.16619
```

I know the probability of being in your `30s` is about 16%.

We'll now find out *P(E)*, which just represents the overall probability of buying something irrespective of your age:

```
PE = float(totalPurchases) / 100000.0
print ("P(Purchase):", PE)

P(Purchase): 0.45012
```

That works out to be, in this example, about 45%. I can just take the total number of things purchased by everybody regardless of age and divide it by the total number of people to get the overall probability of purchase.

Alright, so what do I have here? I have the probability of purchasing something given that you're in your 30s being about 30%, and then I have the probability of purchasing something overall at about 45%.

Now if E and F were independent, if age didn't matter, then I would expect the *P(E|F)* to be about the same as *P(E)*. I would expect the probability of buying something given that you're in your 30s to be about the same as the overall probability of buying something, but they're not, right? And because they're different, that tells me that they are in fact dependent, somehow. So that's a little way of using conditional probability to tease out these dependencies in the data.

Let's do some more notation stuff here. If you see something like *P(E)P(F)* together, that means multiply these probabilities together. I can just take the overall probability of purchase multiplied by the overall probability of being in your `30s`:

```
print ("P(30's)P(Purchase)", PE * PF)

P(30's)P(Purchase) 0.07480544280000001
```

That worked out to about 7.5%.

Just from the way probabilities work, I know that if I want to get the probability of two things happening together, that would be the same thing as multiplying their individual probabilities. So it turns out that *P(E,F)* happening, is the same thing as *P(E)P(F)*.

```
print ("P(30's, Purchase)", float(purchases[30]) / 100000.0)
P(30's, Purchase) 0.04974
```

Now because of the random distribution of data, it doesn't work out to be exactly the same thing. We're talking about probabilities here, remember, but they're in the same ballpark, so that makes sense, about 5% versus 7%, close enough.

Now that is different again from $P(E|F)$, so the probability of both being in your 30s and buying something is different than the probability of buying something given that you're in your 30s.

Now let's just do a little sanity check here. We can check our equation that we saw in the Conditional Probability section earlier, that said that the probability of buying something given that you're in your 30s is the same as the probability of being in your 30s and buying something over the probability of buying something. That is, we check if $P(E|F)=P(E,F)/P(F)$.

```
(float(purchases[30]) / 100000.0) / PF
```

This gives us:

```
Out []:0.29929598652145134
```

Sure enough, it does work out. If I take the probability of buying something given that you're in your 30s over the overall probability, we end up with about 30%, which is pretty much what we came up with originally for $P(E|F)$. So the equation works, yay!

Alright, it's tough to wrap your head around some of this stuff. It's a little bit confusing, I know, but if you need to, go through this again, study it, and make sure you understand what's going on here. I've tried to put in enough examples here to illustrate different combinations of thinking about this stuff. Once you've got it internalized, I'm going to challenge you to actually do a little bit of work yourself here.

Conditional probability assignment

What I want you to do is modify the following Python code which was used in the preceding section.

```
from numpy import random
random.seed(0)

totals = {20:0, 30:0, 40:0, 50:0, 60:0, 70:0}
purchases = {20:0, 30:0, 40:0, 50:0, 60:0, 70:0}
totalPurchases = 0
for _ in range(100000):
    ageDecade = random.choice([20, 30, 40, 50, 60, 70])
    purchaseProbability = 0.4
```

Matplotlib and Advanced Probability Concepts

```
    totals[ageDecade] += 1
    if (random.random() < purchaseProbability):
        totalPurchases += 1
        purchases[ageDecade] += 1
```

Modify it to actually not have a dependency between purchases and age. Make that an evenly distributed chance as well. See what that does to your results. Do you end up with a very different conditional probability of being in your 30s and purchasing something versus the overall probability of purchasing something? What does that tell you about your data and the relationship between those two different attributes? Go ahead and try that, and make sure you can actually get some results from this data and understand what's going on, and I'll run through my own solution to that exercise in just a minute.

So that's conditional probability, both in theory and in practice. You can see there's a lot of little nuances to it and a lot of confusing notation. Go back and go through this section again if you need to wrap your head around it. I gave you a homework assignment, so go off and do that now, see if you can actually modify my code in that IPython Notebook to produce a constant probability of purchase for those different age groups. Come back and we'll take a look at how I solved that problem and what my results were.

My assignment solution

Did you do your homework? I hope so. Let's take a look at my solution to the problem of seeing how conditional probability tells us about whether there's a relationship between age and purchase probability in a fake dataset.

To remind you, what we were trying to do was remove the dependency between age and probability of purchasing and see if we could actually reflect that in our conditional probability values. Here's what I've got:

```
from numpy import random
random.seed(0)

totals = {20:0, 30:0, 40:0, 50:0, 60:0, 70:0}
purchases = {20:0, 30:0, 40:0, 50:0, 60:0, 70:0}
totalPurchases = 0
for _ in range(100000):
    ageDecade = random.choice([20, 30, 40, 50, 60, 70])
    purchaseProbability = 0.4
    totals[ageDecade] += 1
    if (random.random() < purchaseProbability):
        totalPurchases += 1
        purchases[ageDecade] += 1
```

What I've done here is I've taken the original snippet of code for creating our dictionary of age groups and how much was purchased by each age group for a set of 100,000 random people. Instead of making purchase probability dependent on age, I've made it a constant probability of 40%. Now we just have people randomly being assigned to an age group, and they all have the same probability of buying something. Let's go ahead and run that.

Now this time, if I compute the P(E|F), that is, the probability of buying something given that you're in your 30s, I come up with about 40%.

```
PEF = float(purchases[30]) / float(totals[30])
print ("P(purchase | 30s): ", PEF)

P(purchase | 30s):  0.398760454901
```

If I compare that to the overall probability of purchasing, that too is about 40%.

```
PE = float(totalPurchases) / 100000.0
print ("P(Purchase):", PE)

P(Purchase):  0.4003
```

I can see here that the probability of purchasing something given that you're in your 30s is about the same as the probability of purchasing something irrespective of your age (that is, *P(E|F)* is pretty close to *P(E)*). That suggests that there's no real relationship between those two things, and in fact, I know there isn't from this data.

Now in practice, you could just be seeing random chance, so you'd want to look at more than one age group. You'd want to look at more than one data point to see if there really is a relationship or not, but this is an indication that there's no relationship between age and probability of purchase in this sample data that we modified.

So, that's conditional probability in action. Hopefully your solution was fairly close and had similar results. If not, go back and study my solution. It's right there in the data files for this book, ConditionalProbabilitySolution.ipynb, if you need to open it up and study it and play around with it. Obviously, the random nature of the data will make your results a little bit different and will depend on what choice you made for the overall purchase probability, but that's the idea.

And with that behind us, let's move on to Bayes' theorem.

Bayes' theorem

Now that you understand conditional probability, you can understand how to apply Bayes' theorem, which is based on conditional probability. It's a very important concept, especially if you're going into the medical field, but it is broadly applicable too, and you'll see why in a minute.

You'll hear about this a lot, but not many people really understand what it means or its significance. It can tell you very quantitatively sometimes when people are misleading you with statistics, so let's see how that works.

First, let's talk about Bayes' theorem at a high level. Bayes' theorem is simply this: the probability of A given B is equal to the probability of A times the probability of B given A over the probability of B. So you can substitute A and B with whatever you want.

$$P(A|B) = \frac{P(A)P(B|A)}{P(B)}$$

The key insight is that the probability of something that depends on B depends very much on the base probability of B and A. People ignore this all the time.

One common example is drug testing. We might say, what's the probability of being an actual user of a drug given that you tested positive for it. The reason Bayes' theorem is important is that it calls out that this very much depends on both the probability of A and the probability of B. The probability of being a drug user given that you tested positive depends very much on the base overall probability of being a drug user and the overall probability of testing positive. The probability of a drug test being accurate depends a lot on the overall probability of being a drug user in the population, not just the accuracy of the test.

It also means that the probability of B given A is not the same thing as the probability of A given B. That is, the probability of being a drug user given that you tested positive can be very different from the probability of testing positive given that you're a drug user. You can see where this is going. That is a very real problem where diagnostic tests in medicine or drug tests yield a lot of false positives. You can still say that the probability of a test detecting a user can be very high, but it doesn't necessarily mean that the probability of being a user given that you tested positive is high. Those are two different things, and Bayes' theorem allows you to quantify that difference.

Let's nail that example home a little bit more.

Again, a drug test can be a common example of applying Bayes' theorem to prove a point. Even a highly accurate drug test can produce more false positives than true positives. So in our example here, we're going to come up with a drug test that can accurately identify users of a drug 99% of the time and accurately has a negative result for 99% of non-users, but only 0.3% of the overall population actually uses the drug in question. So we have a very small probability of actually being a user of a drug. What seems like a very high accuracy of 99% isn't actually high enough, right?

We can work out the math as follows:

- Event A = is a user of the drug
- Event B = tested positively for the drug

So let event A mean that you're a user of some drug, and event B the event that you tested positively for the drug using this drug test.

We need to work out the probability of testing positively overall. We can work that out by taking the sum of probability of testing positive if you are a user and the probability of testing positive if you're not a user. So, P(B) works out to 1.3% (0.99*0.003+0.01*0.997) in this example. So we have a probability of B, the probability of testing positively for the drug overall without knowing anything else about you.

Let's do the math and calculate the probability of being a user of the drug given that you tested positively.

$$P(A|B) = \frac{P(A)P(B|A)}{P(B)} = \frac{0.003 * 0.99}{0.013} = 22.8\%$$

So the probability of a positive test result given that you're actually a drug user works out as the probability of being a user of the drug overall *(P(A))*, which is 3% (you know that 3% of the population is a drug user) multiplied by *P(B|A)* that is the probability of testing positively given that you're a user divided by the probability of testing positively overall which is 1.3%. Again, this test has what sounds like a very high accuracy of 99%. We have 0.3% of the population which uses a drug multiplied by the accuracy of 99% divided by the probability of testing positively overall, which is 1.3%. So the probability of being an actual user of this drug given that you tested positive for it is only 22.8%. So even though this drug test is accurate 99% of the time, it's still providing a false result in most of the cases where you're testing positive.

Even though P(B|A) is high (99%), it doesn't mean P(A|B) is high.

People overlook this all the time, so if there's one lesson to be learned from Bayes' theorem, it is to always take these sorts of things with a grain of salt. Apply Bayes' theorem to these actual problems and you'll often find that what sounds like a high accuracy rate can actually be yielding very misleading results if you're dealing with a low overall incidence of a given problem. We see the same thing in cancer screening and other sorts of medical screening as well. That's a very real problem; there's a lot of people getting very, very real and very unnecessary surgery as a result of not understanding Bayes' theorem. If you're going into the medical profession with big data, please, please, please remember this theorem.

So that's Bayes' theorem. Always remember that the probability of something given something else is not the same thing as the other way around, and it actually depends a lot on the base probabilities of both of those two things that you're measuring. It's a very important thing to keep in mind, and always look at your results with that in mind. Bayes' theorem gives you the tools to quantify that effect. I hope it proves useful.

Summary

In this chapter, we talked about plotting and graphing your data and how to make your graphs look pretty using the `matplotlib` library in Python. We also walked through the concepts of covariance and correlation. We looked at some examples and figured out covariance and correlation using Python. We analyzed the concept of conditional probability and saw some examples to understand it better. Finally, we saw Bayes' theorem and its importance, especially in the medical field.

In the next chapter, we'll talk about predictive models.

4
Predictive Models

In this chapter, we're going to look at what predictive modeling is and how it uses statistics to predict outcomes from existing data. We'll cover real world examples to understand the concepts better. We'll see what regression analysis means and analyze some of its forms in detail. We'll also look at an example which predicts the price of a car for us.

These are the topics that we'll cover in this chapter:

- Linear regression and how to implement it in Python
- Polynomial regression, its application and examples
- Multivariate regression and how to implement it in Python
- An example we'll build that predicts the price of a car using Python
- The concept of multi-level models and some things to know about them

Linear regression

Let's talk about regression analysis, a very popular topic in data science and statistics. It's all about trying to fit a curve or some sort of function, to a set of observations and then using that function to predict new values that you haven't seen yet. That's all there is to linear regression!

Predictive Models

So, linear regression is fitting a straight line to a set of observations. For example, let's say that I have a bunch of people that I measured and the two features that I measured of these people are their weight and their height:

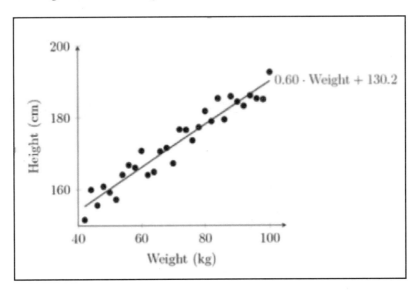

I'm showing the weight on the *x*-axis and the height on the *y*-axis, and I can plot all these data points, as in the people's weight versus their height, and I can say, "Hmm, that looks like a linear relationship, doesn't it? Maybe I can fit a straight line to it and use that to predict new values", and that's what linear regression does. In this example, I end up with a slope of 0.6 and a *y*-intercept of 130.2 which define a straight line (the equation of a straight line is $y=mx+b$, where m is the slope and b is the *y*-intercept). Given a slope and a *y*-intercept, that fits the data that I have best, I can use that line to predict new values.

You can see that the weights that I observed only went up to people that weighed 100 kilograms. What if I had someone who weighed 120 kilograms? Well, I could use that line to then figure out where would the height be for someone with 120 kilograms based on this previous data.

I don't know why they call it regression. Regression kind of implies that you're doing something backwards. I guess you can think of it in terms of you're creating a line to predict new values based on observations you made in the past, backwards in time, but it seems like a little bit of a stretch. It's just a confusing term quite honestly, and one way that we kind of obscure what we do with very simple concepts using very fancy terminology. All it is, is fitting a straight line to a set of data points.

Chapter 4

The ordinary least squares technique

How does linear regression work? Well internally, it uses a technique called ordinary least squares; it's also known as, OLS. You might see that term tossed around as well. The way it works is it tries to minimize the squared error between each point and the line, where the error is just the distance between each point and the line that you have.

So, we sum up all the squares of those errors, which sounds a lot like when we computed variance, right, except that instead of relative to the mean, it's relative to the line that we're defining. We can measure the variance of the data points from that line, and by minimizing that variance, we can find the line that fits it the best:

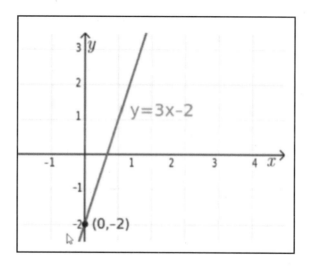

Now you'll never have to actually do this yourself the hard way, but if you did have to for some reason, or if you're just curious about what happens under the hood, I'll now describe the overall algorithm for you and how you would actually go about computing the slope and y-intercept yourself the hard way if you need to one day. It's really not that complicated.

Remember the slope-intercept equation of a line? It is $y=mx+c$. The slope just turns out to be the correlation between the two variables times the standard deviation in Y divided by the standard deviation in X. It might seem a little bit weird that standard deviation just kind of creeps into the math naturally there, but remember correlation had standard deviation baked into it as well, so it's not too surprising that you have to reintroduce that term.

[137]

The intercept can then be computed as the mean of the Y minus the slope times the mean of X. Again, even though that's really not that difficult, Python will do it all for you, but the point is that these aren't complicated things to run. They can actually be done very efficiently.

Remember that least squares minimize the sum of squared errors from each point to the line. Another way of thinking about linear regression is that you're defining a line that represents the maximum likelihood of an observation line there; that is, the maximum probability of the y value being something for a given x value.

People sometimes call linear regression maximum likelihood estimation, and it's just another example of people giving a fancy name to something that's very simple, so if you hear someone talk about maximum likelihood estimation, they're really talking about regression. They're just trying to sound really smart. But now you know that term too, so you too can sound smart.

The gradient descent technique

There is more than one way to do linear regression. We've talked about ordinary least squares as being a simple way of fitting a line to a set of data, but there are other techniques as well, gradient descent being one of them, and it works best in three-dimensional data. So, it tries to follow the contours of the data for you. It's very fancy and obviously a little bit more computationally expensive, but Python does make it easy for you to try it out if you want to compare it to ordinary least squares.

Using the gradient descent technique can make sense when dealing with 3D data.

Usually though, least squares is a perfectly good choice for doing linear regression, and it's always a legitimate thing to do, but if you do run into gradient descent, you will know that that is just an alternate way of doing linear regression, and it's usually seen in higher dimensional data.

The co-efficient of determination or r-squared

So how do I know how good my regression is? How well does my line fit my data? That's where r-squared comes in, and r-squared is also known as the coefficient of determination. Again, someone trying to sound smart might call it that, but usually it's called r-squared.

It is the fraction of the total variation in Y that is captured by your models. So how well does your line follow that variation that's happening? Are we getting an equal amount of variance on either side of your line or not? That's what r-squared is measuring.

Computing r-squared

To actually compute the value, take 1 minus the sum of the squared errors over the sum of the squared variations from the mean:

$$1.0 - \frac{sum\ of\ squared\ errors}{sum\ of\ squared\ variation\ from\ mean}$$

So, it's not very difficult to compute, but again, Python will give you functions that will just compute that for you, so you'll never have to actually do that math yourself.

Interpreting r-squared

For r-squared, you will get a value that ranges from 0 to 1. Now 0 means your fit is terrible. It doesn't capture any of the variance in your data. While 1 is a perfect fit, where all of the variance in your data gets captured by this line, and all of the variance you see on either side of your line should be the same in that case. So 0 is bad, and 1 is good. That's all you really need to know. Something in between is something in between. A low r-squared value means it's a poor fit, a high r-squared value means it's a good fit.

As you'll see in the coming sections, there's more than one way to do regression. Linear regression is one of them. It's a very simple technique, but there are other techniques as well, and you can use r-squared as a quantitative measure of how good a given regression is to a set of data points, and then use that to choose the model that best fits your data.

Predictive Models

Computing linear regression and r-squared using Python

Let's now play with linear regression and actually compute some linear regression and r-squared. We can start by creating a little bit of Python code here that generates some *random-ish* data that is in fact linearly correlated.

In this example I'm going to fake some data about page rendering speeds and how much people purchase, just like a previous example. We're going to fabricate a linear relationship between the amount of time it takes for a website to load and the amount of money people spend on that website:

```
%matplotlib inline
import numpy as np
from pylab import *
pageSpeeds = np.random.normal(3.0, 1.0, 1000)
purchaseAmount = 100 - (pageSpeeds + np.random.normal(0, 0.1,
1000)) * 3
scatter(pageSpeeds, purchaseAmount)
```

All I've done here is I've made a random, a normal distribution of page speeds centered around 3 seconds with a standard deviation of 1 second. I've made the purchase amount a linear function of that. So, I'm making it 100 minus the page speeds plus some normal random distribution around it, times 3. And if we scatter that, we can see that the data ends up looking like this:

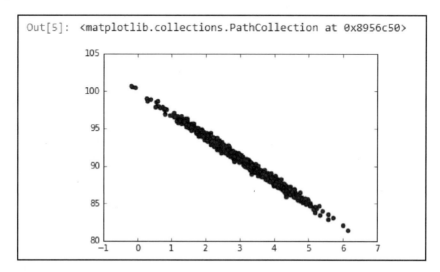

You can see just by eyeballing it that there's definitely a linear relationship going on there, and that's because we did hardcode a real linear relationship in our source data.

Now let's see if we can tease that out and find the best fit line using ordinary least squares. We talked about how to do ordinary least squares and linear regression, but you don't have to do any of that math yourself because the SciPy package has a `stats` package that you can import:

```
from scipy import stats
slope, intercept, r_value, p_value, std_err = stats.linregress(pageSpeeds, purchaseAmount)
```

You can import `stats` from `scipy`, and then you can just call `stats.linregress()` on your two features. So, we have a list of page speeds (`pageSpeeds`) and a corresponding list of purchase amounts (`purchaseAmount`). The `linregress()` function will give us back a bunch of stuff, including the slope, the intercept, which is what I need to define my best fit line. It also gives us the `r_value`, from which we can get r-squared to measure the quality of that fit, and a couple of things that we'll talk about later on. For now, we just need slope, intercept, and `r_value`, so let's go ahead and run these. We'll begin by finding the linear regression best fit:

```
r_value ** 2
```

This is what your output should look like:

```
Out[4]: 0.98984146047689425
```

Now the r-squared value of the line that we got back is 0.99, that's almost 1.0. That means we have a really good fit, which isn't too surprising because we made sure there was a real linear relationship between this data. Even though there is some variance around that line, our line captures that variance. We have roughly the same amount of variance on either side of the line, which is a good thing. It tells us that we do have a linear relationship and our model is a good fit for the data that we have.

Let's plot that line:

```
import matplotlib.pyplot as plt
def predict(x):
return slope * x + intercept
fitLine = predict(pageSpeeds)
plt.scatter(pageSpeeds, purchaseAmount)
plt.plot(pageSpeeds, fitLine, c='r')
plt.show()
```

Predictive Models

The following is the output to the preceding code:

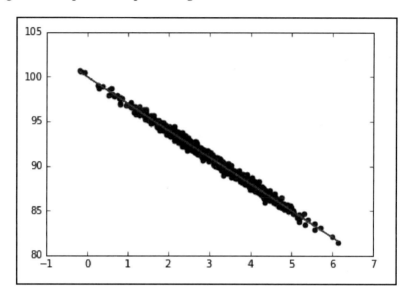

This little bit of code will create a function to draw the best fit line alongside the data. There's a little bit more Matplotlib magic going on here. We're going to make a `fitLine` list and we're going to use the `predict()` function we wrote to take the `pageSpeeds`, which is our *x*-axis, and create the Y function from that. So instead of taking the observations for amount spent, we're going to find the predicted ones just using the `slope` times x plus the `intercept` that we got back from the `linregress()` call above. Essentially here, we're going to do a scatter plot like we did before to show the raw data points, which are the observations.

Then we're also going to call `plot` on that same `pyplot` instance using our `fitLine` that we created using the line equation that we got back, and show them all both together. When we do that, it looks like the following graph:

Chapter 4

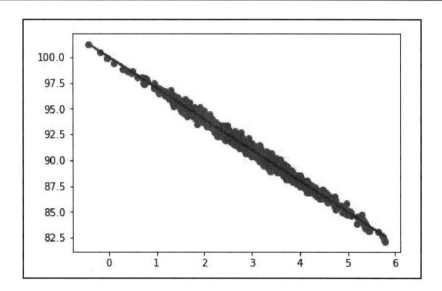

You can see that our line is in fact a great fit for our data! It goes right smack down the middle, and all you need to predict new values is this predict function. Given a new previously unseen page speed, we could predict the amount spent just using the slope times the page speed plus the intercept. That's all there is to it, and I think it's great!

Activity for linear regression

Time now to get your hands dirty. Try increasing the random variation in the test data and see if that has any impact. Remember, the r-squared is a measure of the fit, of how much do we capture the variance, so the amount of variance, well... why don't you see if it actually makes a difference or not.

That's linear regression, a pretty simple concept. All we're doing is fitting a straight line to set of observations, and then we can use that line to make predictions of new values. That's all there is to it. But why limit yourself to a line? There's other types of regression we can do that are more complex. We'll explore these next.

Polynomial regression

We've talked about linear regression where we fit a straight line to a set of observations. Polynomial regression is our next topic, and that's using higher order polynomials to fit your data. So, sometimes your data might not really be appropriate for a straight line. That's where polynomial regression comes in.

Polynomial regression is a more general case of regression. So why limit yourself to a straight line? Maybe your data doesn't actually have a linear relationship, or maybe there's some sort of a curve to it, right? That happens pretty frequently.

Not all relationships are linear, but the linear regression is just one example of a whole class of regressions that we can do. If you remember the linear regression line that we ended up with was of the form $y = mx + b$, where we got back the values m and b from our linear regression analysis from ordinary least squares, or whatever method you choose. Now this is just a first order or a first-degree polynomial. The order or the degree is the power of x that you see. So that's the first-order polynomial.

Now if we wanted, we could also use a second-order polynomial, which would look like $y = ax^2 + bx + c$. If we were doing a regression using a second-order polynomial, we would get back values for a, b, and c. Or we could do a third-order polynomial that has the form $ax^3 + bx^2 + cx + d$. The higher the orders get, the more complex the curves you can represent. So, the more powers of x you have blended together, the more complicated shapes and relationships you can get.

But more degrees aren't always better. Usually there's some natural relationship in your data that isn't really all that complicated, and if you find yourself throwing very large degrees at fitting your data, you might be overfitting!

Beware of overfitting!

- Don't use more degrees than you need
- Visualize your data first to see how complex of a curve there might really be
- Visualize the fit and check if your curve going out of its way to accommodate outliers
- A high r-squared simply means your curve fits your training data well; it may or may not be good predictor

If you have data that's kind of all over the place and has a lot of variance, you can go crazy and create a line that just like goes up and down to try to fit that data as closely as it can, but in fact that doesn't represent the intrinsic relationship of that data. It doesn't do a good job of predicting new values.

So always start by just visualizing your data and think about how complicated does the curve really needs to be. Now you can use r-squared to measure how good your fit is, but remember, that's just measuring how well this curve fits your training data—that is, the data that you're using to actually make your predictions based off of. It doesn't measure your ability to predict accurately going forward.

Later, we'll talk about some techniques for preventing overfitting called **train/test**, but for now you're just going to have to eyeball it to make sure that you're not overfitting and throwing more degrees at a function than you need to. This will make more sense when we explore an example, so let's do that next.

Implementing polynomial regression using NumPy

Fortunately, NumPy has a `polyfit` function that makes it super easy to play with this and experiment with different results, so let's go take a look. Time for fun with polynomial regression. I really do think it's fun, by the way. It's kind of cool seeing all that high school math actually coming into some practical application. Go ahead and open the `PolynomialRegression.ipynb` and let's have some fun.

Let's create a new relationship between our page speeds, and our purchase amount fake data, and this time we're going to create a more complex relationship that's not linear. We're going to take the page speed and make it some function of the division of page speed for the purchase amount:

```
%matplotlib inline
from pylab import *
np.random.seed(2)
pageSpeeds = np.random.normal(3.0, 1.0, 1000)
purchaseAmount = np.random.normal(50.0, 10.0, 1000) / pageSpeeds
scatter(pageSpeeds, purchaseAmount)
```

Predictive Models

If we do a scatter plot, we end up with the following:

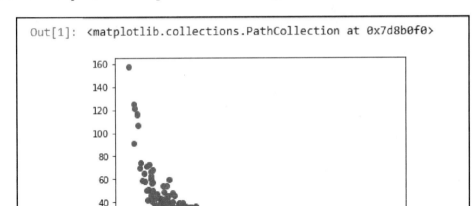

By the way, if you're wondering what the `np.random.seed` line does, it creates a random seed value, and it means that when we do subsequent random operations they will be deterministic. By doing this we can make sure that, every time we run this bit of code, we end up with the same exact results. That's going to be important later on because I'm going to suggest that you come back and actually try different fits to this data to compare the fits that you get. So, it's important that you're starting with the same initial set of points.

You can see that that's not really a linear relationship. We could try to fit a line to it and it would be okay for a lot of the data, maybe down at the right side of the graph, but not so much towards the left. We really have more of an exponential curve.

Now it just happens that NumPy has a `polyfit()` function that allows you to fit any degree polynomial you want to this data. So, for example, we could say our *x*-axis is an array of the page speeds (`pageSpeeds`) that we have, and our *y*-axis is an array of the purchase amounts (`purchaseAmount`) that we have. We can then just call `np.polyfit(x, y, 4)`, meaning that we want a fourth degree polynomial fit to this data.

```
x = np.array(pageSpeeds)
y = np.array(purchaseAmount)
p4 = np.poly1d(np.polyfit(x, y, 4))
```

Let's go ahead and run that. It runs pretty quickly, and we can then plot that. So, we're going to create a little graph here that plots our scatter plot of original points versus our predicted points.

```
import matplotlib.pyplot as plt
xp = np.linspace(0, 7, 100)
plt.scatter(x, y)
plt.plot(xp, p4(xp), c='r')
plt.show()
```

The output looks like the following graph:

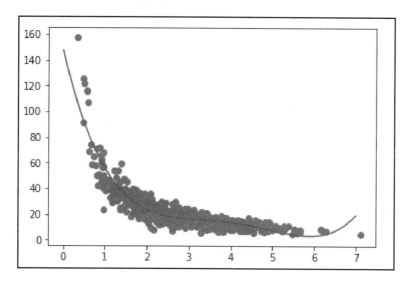

At this point, it looks like a reasonably good fit. What you want to ask yourself though is, "Am I overfitting? Does my curve look like it's actually going out of its way to accommodate outliers?" I find that that's not really happening. I don't really see a whole lot of craziness going on.

If I had a really high order polynomial, it might swoop up at the top to catch that one outlier and then swoop downwards to catch the outliers there, and get a little bit more stable through where we have a lot of density, and maybe then it could potentially go all over the place trying to fit the last set of outliers at the end. If you see that sort of nonsense, you know you have too many orders, too many degrees in your polynomial, and you should probably bring it back down because, although it fits the data that you observed, it's not going to be useful for predicting data you haven't seen.

Predictive Models

Imagine I have some curve that swoops way up and then back down again to fit outliers. My prediction for something in between there isn't going to be accurate. The curve really should be in the middle. Later in this book we'll talk about the main ways of detecting such overfitting, but for now, please just observe it and know we'll go deeper later.

Computing the r-squared error

Now we can measure the r-squared error. By taking the y and the predicted values (p4(x)) in the r2_score() function that we have in sklearn.metrics, we can compute that.

```
from sklearn.metrics import r2_score
r2 = r2_score(y, p4(x))
print r2
```

The output is as follows:

```
0.82937663963
```

Our code compares a set of observations to a set of predictions and computes r-squared for you, and with just one line of code! Our r-squared for this turns out to be 0.829, which isn't too bad. Remember, zero is bad, one is good. 0.82 is to pretty close to one, not perfect, and intuitively, that makes sense. You can see that our line is pretty good in the middle section of the data, but not so good out at the extreme left and not so good down at the extreme right. So, 0.82 sounds about right.

Activity for polynomial regression

I recommend that you get down and dirty with this stuff. Try different orders of polynomials. Go back up to where we ran the polyfit() function and try different values there besides 4. You can use 1, and that would go back to a linear regression, or you could try some really high amount like 8, and maybe you'll start to see overfitting. So see what effect that has. You're going to want to change that. For example, let's go to a third-degree polynomial.

```
x = np.array(pageSpeeds)
y = np.array(purchaseAmount)
p4 = np.poly1d(np.polyfit(x, y, 3))
```

Just keep hitting run to go through each step and you can see the it's effect as...

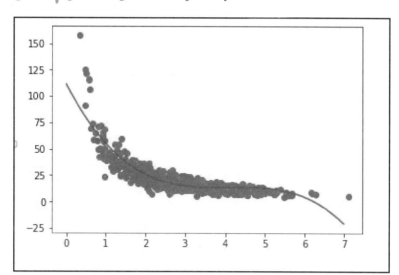

Our third-degree polynomial is definitely not as good a fit as the fourth-degree polynomial. If you actually measure the r-squared error, it would actually turn out worse, quantitatively; but if I go too high, you might start to see overfitting. So just have some fun with that, play around different values, and get a sense of what different orders of polynomials do to your regression. Go get your hands dirty and try to learn something.

So that's polynomial regression. Again, you need to make sure that you don't put more degrees at the problem than you need to. Use just the right amount to find what looks like an intuitive fit to your data. Too many can lead to overfitting, while too few can lead to a poor fit... so you can use both your eyeballs for now, and the r-squared metric, to figure out what the right number of degrees are for your data. Let's move on.

Multivariate regression and predicting car prices

What happens then, if we're trying to predict some value that is based on more than one other attribute? Let's say that the height of people not only depends on their weight, but also on their genetics or some other things that might factor into it. Well, that's where multivariate analysis comes in. You can actually build regression models that take more than one factor into account at once. It's actually pretty easy to do with Python.

Let's talk about multivariate regression, which is a little bit more complicated. The idea of multivariate regression is this: what if there's more than one factor that influences the thing you're trying to predict?

In our previous examples, we looked at linear regression. We talked about predicting people's heights based on their weight, for example. We assumed that the weight was the only thing that influenced their height, but maybe there are other factors too. We also looked at the effect of page speed on purchase amounts. Maybe there's more that influences purchase amounts than just page speed, and we want to find how these different factors all combine together to influence that value. So that's where multivariate regression comes in.

The example we're going to look at now is as follows. Let's say that you're trying to predict the price that a car will sell for. It might be based on many different features of that car, such as the body style, the brand, the mileage; who knows, even on how good the tires are. Some of those features are going to be more important than others toward predicting the price of a car, but you want to take into account all of them at once.

So our way forwards here is still going to use the least-squares approach to fit a model to your set of observations. The difference is that we're going to have a bunch of coefficients for each different feature that you have.

So, for example, the price model that we end up with might be a linear relationship of alpha, some constant, kind of like your y-intercept was, plus some coefficient of the mileage, plus some coefficient of the age, plus some coefficient of how many doors it has:

$$\text{For example, } price = \alpha + \beta_1 \text{mileage} + \beta_2 \text{age} + \beta_2 \text{doors}$$

Once you end up with those coefficients, from least squares analysis, we can use that information to figure out, well, how important are each of these features to my model. So, if I end up with a very small coefficient for something like the number of doors, that implies that the number of doors isn't that important, and maybe I should just remove it from my model entirely to keep it simpler.

This is something that I really should say more often in this book. You always want to do the simplest thing that works in data science. Don't over complicate things, because it's usually the simple models that work the best. If you can find just the right amount of complexity, but no more, that's usually the right model to go with. Anyway, those coefficients give you a way of actually, "Hey some of these things are more important than others. Maybe I can discard some of these factors."

Now we can still measure the quality of a fit with multivariate regression using r-squared. It works the same way, although one thing you need to assume when you're doing multivariate regression is that the factors themselves are not dependent on each other... and that's not always true. So sometimes you need to keep that little caveat in the back of your head. For example, in this model we're going to assume that mileage and age of the car are not related; but in fact, they're probably pretty tightly related! This is a limitation of this technique, and it might not be capturing an effect at all.

Multivariate regression using Python

Fortunately there's a `statsmodel` package available for Python that makes doing multivariate regression pretty easy. Let's just dive in and see how it works. Let's do some multivariate regression using Python. We're going to use some real data here about car values from the Kelley Blue Book.

```
import pandas as pd
df = pd.read_excel('http://cdn.sundog-soft.com/Udemy/DataScience/cars.xls')
```

We're going to introduce a new package here called `pandas`, which lets us deal with tabular data really easily. It lets us read in tables of data and rearrange them, and modify them, and slice them and dice them in different ways. We're going to be using that a lot going forward.

We're going to import `pandas` as `pd`, and `pd` has a `read_Excel()` function that we can use to go ahead and read a Microsoft Excel spreadsheet from the Web through HTTP. So, pretty awesome capabilities of pandas there.

I've gone ahead and hosted that file for you on my own domain, and if we run that, it will load it into what's called a `DataFrame` object that we're referring to as `df`. Now I can call `head()` on this `DataFrame` to just show the first few lines of it:

```
df.head()
```

Predictive Models

The following is the output for the preceding code:

Out[2]:

	Price	Mileage	Make	Model	Trim	Type	Cylinder	Liter	Doors	Cruise	Sound	Leather
0	17314.103129	8221	Buick	Century	Sedan 4D	Sedan	6	3.1	4	1	1	1
1	17542.036083	9135	Buick	Century	Sedan 4D	Sedan	6	3.1	4	1	1	0
2	16218.847862	13196	Buick	Century	Sedan 4D	Sedan	6	3.1	4	1	1	0
3	16336.913140	16342	Buick	Century	Sedan 4D	Sedan	6	3.1	4	1	0	0
4	16339.170324	19832	Buick	Century	Sedan 4D	Sedan	6	3.1	4	1	0	1

The actual dataset is much larger. This is just the first few samples. So, this is real data of mileage, make, model, trim, type, doors, cruise, sound and leather.

OK, now we're going to use `pandas` to split that up into the features that we care about. We're going to create a model that tries to predict the price just based on the mileage, the model, and the number of doors, and nothing else.

```
import statsmodels.api as sm

df['Model_ord'] = pd.Categorical(df.Model).codes
X = df[['Mileage', 'Model_ord', 'Doors']]
y = df[['Price']]

X1 = sm.add_constant(X)
est = sm.OLS(y, X1).fit()

est.summary()
```

Now the problem that I run into is that the model is a text, like Century for Buick, and as you recall, everything needs to be a number when I'm doing this sort of analysis. In the code, I use this `Categorical()` function in `pandas` to actually convert the set of model names that it sees in the `DataFrame` into a set of numbers; that is, a set of codes. I'm going to say my input for this model on the x-axis is mileage (`Mileage`), model converted to an ordinal value (`Model_ord`), and the number of doors (`Doors`). What I'm trying to predict on the y-axis is the price (`Price`).

The next two lines of the code just create a model that I'm calling `est` that uses ordinary least squares, OLS, and fits that using the columns that I give it, `Mileage`, `Model_ord`, and `Doors`. Then I can use the summary call to print out what my model looks like:

Out[3]:

OLS Regression Results

Dep. Variable:	Price	R-squared:	0.042
Model:	OLS	Adj. R-squared:	0.038
Method:	Least Squares	F-statistic:	11.57
Date:	Tue, 26 Jan 2016	Prob (F-statistic):	1.98e-07
Time:	12:18:05	Log-Likelihood:	-8519.1
No. Observations:	804	AIC:	1.705e+04
Df Residuals:	800	BIC:	1.706e+04
Df Model:	3		
Covariance Type:	nonrobust		

	coef	std err	t	P>|t|	[95.0% Conf. Int.]	
const	3.125e+04	1809.549	17.272	0.000	2.77e+04	3.48e+04
Mileage	-0.1765	0.042	-4.227	0.000	-0.259	-0.095
Model_ord	-39.0387	39.326	-0.993	0.321	-116.234	38.157
Doors	-1652.9303	402.649	-4.105	0.000	-2443.303	-862.558

Omnibus:	206.410	Durbin-Watson:	0.080
Prob(Omnibus):	0.000	Jarque-Bera (JB):	470.872
Skew:	1.379	Prob(JB):	5.64e-103
Kurtosis:	5.541	Cond. No.	1.15e+05

You can see here that the r-squared is pretty low. It's not that good of a model, really, but we can get some insight into what the various errors are, and interestingly, the lowest standard error is associated with the mileage.

Predictive Models

Now I have said before that the coefficient is a way of determining which items matter, and that's only true though if your input data is normalized. That is, if everything's on the same scale of 0 to 1. If it's not, then these coefficients are kind of compensating for the scale of the data that it's seeing. If you're not dealing with normalized data, as in this case, it's more useful to look at the standard errors. In this case, we can see that the mileage is actually the biggest factor of this particular model. Could we have figured that out earlier? Well, we could have just done a little bit of slicing and dicing to figure out that the number of doors doesn't actually influence the price much at all. Let's run the following little line:

```
y.groupby(df.Doors).mean()
```

A little bit of `pandas` syntax there. It's pretty cool that you can do it in Python in one line of code! That will print out a new `DataFrame` that shows the mean price for the given number of doors:

Out[5]:		Price
	Doors	
	2	23807.135520
	4	20580.670749

I can see the average two-door car sells for actually more than the average four-door car. If anything there's a negative correlation between number of doors and price, which is a little bit surprising. This is a small dataset, though, so we can't read a whole lot of meaning into it of course.

Activity for multivariate regression

As an activity, please mess around with the fake input data where you want. You can download the data and mess around with the spreadsheet. Read it from your local hard drive instead of from HTTP, and see what kind of differences you can have. Maybe you can fabricate a dataset that has a different behavior and has a better model that fits it. Maybe you can make a wiser choice of features to base your model off of. So, feel free to mess around with that and let's move on.

There you have it: multivariate analysis and an example of it running. Just as important as the concept of multivariate analysis, which we explored, was some of the stuff that we did in that Python notebook. So, you might want to go back there and study exactly what's going on.

We introduced pandas and the way to work with pandas and DataFrame objects. pandas a very powerful tool. We'll use it more in future sections, but make sure you're starting to take notice of these things because these are going to be important techniques in your Python skills for managing large amounts of data and organizing your data.

Multi-level models

It makes sense now to talk about multi-level models. This is definitely an advanced topic, and I'm not going to get into a whole lot of detail here. My objective right now is to introduce the concept of multi-level models to you, and let you understand some of the challenges and how to think about them when you're putting them together. That's it.

The concept of multi-level models is that some effects happen at various levels in the hierarchy. For example, your health. Your health might depend on how healthy your individual cells are, and those cells might be a function of how healthy the organs that they're inside are, and the health of your organs might depend on the health of you as a whole. Your health might depend in part on your family's health and the environment your family gives you. And your family's health in turn might depend on some factors of the city that you live in, how much crime is there, how much stress is there, how much pollution is there. And even beyond that, it might depend on factors in the entire world that we live in. Maybe just the state of medical technology in the world is a factor, right?

Another example: your wealth. How much money do you make? Well, that's a factor of your individual hard work, but it's also a factor of the work that your parents did, how much money were they able to invest into your education and the environment that you grew up in, and in turn, how about your grandparents? What sort of environment were they able to create and what sort of education were they able to offer for your parents, which in turn influenced the resources they have available for your own education and upbringing.

These are all examples of multi-level models where there is a hierarchy of effects that influence each other at larger and larger scales. Now the challenge of multi-level models is to try to figure out, "Well, how do I model these interdependencies? How do I model all these different effects and how they affect each other?"

The challenge here is to identify the factors in each level that actually affect the thing you're trying to predict. If I'm trying to predict overall SAT scores, for example, I know that depends in part on the individual child that's taking the test, but what is it about the child that matters? Well, it might be the genetics, it might be their individual health, the individual brain size that they have. You can think of any number of factors that affect the individual that might affect their SAT score. And then if you go up another level, look at their home environment, look at their family. What is it about their families that might affect their SAT scores? How much education were they able to offer? Are the parents able to actually tutor the children in the topics that are on the SAT? These are all factors at that second level that might be important. What about their neighborhood? The crime rate of the neighborhood might be important. The facilities they have for teenagers and keeping them off the streets, things like that.

The point is you want to keep looking at these higher levels, but at each level identify the factors that impact the thing you're trying to predict. I can keep going up to the quality of the teachers in their school, the funding of the school district, the education policies at the state level. You can see there are different factors at different levels that all feed into this thing you're trying to predict, and some of these factors might exist at more than one level. Crime rate, for example, exists at the local and state levels. You need to figure out how those all interplay with each other as well when you're doing multi-level modeling.

As you can imagine, this gets very hard and very complicated very quickly. It is really way beyond the scope of this book, or any introductory book in data science. This is hard stuff. There are entire thick books about it, you could do an entire book about it that would be a very advanced topic.

So why am I even mentioning multi-level models? It is because I've seen it mentioned on job descriptions, in a couple of cases, as something that they want you to know about in a couple of cases. I've never had to use it in practice, but I think the important thing from the standpoint of getting a career in data science is that you at least are familiar with the concept, and you know what it means and some of the challenges involved in creating a multi-level model. I hope I've given you those concepts. With that, we can move on to the next section.

There you have the concepts of multi-level models. It's a very advanced topic, but you need to understand what the concept is, at least, and the concept itself is pretty simple. You just are looking at the effects at different levels, different hierarchies when you're trying to make a prediction. So maybe there are different layers of effects that have impacts on each other, and those different layers might have factors that interrelate with each other as well. Multi-level modeling tries to take account of all those different hierarchies and factors and how they interplay with each other. Rest assured that's all you need to know for now.

Summary

In this chapter, we talked about regression analysis, which is trying to fit a curve to a set of training data and then using it to predict new values. We saw its different forms. We looked at the concept of linear regression and its implementation in Python.

We learned what polynomial regression is, that is, using higher degree polynomials to create better, complex curves for multi-dimensional data. We also saw its implementation in Python.

We then talked about multivariate regression, which is a little bit more complicated. We saw how it is used when there are multiple factors affecting the data that we're predicting. We looked at an interesting example, which predicts the price of a car using Python and a very powerful tool, pandas.

Finally, we looked at the concept of multi-level models. We understood some of the challenges and how to think about them when you're putting them together. In the next chapter, we'll learn some machine learning techniques using Python.

5
Machine Learning with Python

In this chapter, we get into machine learning and how to actually implement machine learning models in Python.

We'll examine what supervised and unsupervised learning means, and how they're different from each other. We'll see techniques to prevent overfitting, and then look at an interesting example where we implement a spam classifier. We'll analyze what K-Means clustering is a long the way, with a working example that clusters people based on their income and age using scikit-learn!

We'll also cover a really interesting application of machine learning called **decision trees** and we'll build a working example in Python that predict shiring decisions in a company. Finally, we'll walk through the fascinating concepts of ensemble learning and SVMs, which are some of my favourite machine learning areas!

More specifically, we'll cover the following topics:

- Supervised and unsupervised learning
- Avoiding overfitting by using train/test
- Bayesian methods
- Implementation of an e-mail spam classifier with Naïve Bayes
- Concept of K-means clustering
- Example of clustering in Python
- Entropy and how to measure it
- Concept of decision trees and its example in Python
- What is ensemble learning
- **Support Vector Machine (SVM)** and its example using scikit-learn

Machine learning and train/test

So what is machine learning? Well, if you look it up on Wikipedia or whatever, it'll say that it is algorithms that can learn from observational data and can make predictions based on it. It sounds really fancy, right? Like artificial intelligence stuff, like you have a throbbing brain inside of your computer. But in reality, these techniques are usually very simple.

We've already looked at regressions, where we took a set of observational data, we fitted a line to it, and we used that line to make predictions. So by our new definition, that was machine learning! And your brain works that way too.

Another fundamental concept in machine learning is something called **train/test**, which lets us very cleverly evaluate how good a machine learning model we've made. As we look now at unsupervised and supervised learning, you'll see why train/test is so important to machine learning.

Unsupervised learning

Let's talk in detail now about two different types of machine learning: supervised and unsupervised learning. Sometimes there can be kind of a blurry line between the two, but the basic definition of unsupervised learning is that you're not giving your model any answers to learn from. You're just presenting it with a group of data and your machine learning algorithm tries to make sense out of it given no additional information:

Let's say I give it a bunch of different objects, like these balls and cubes and sets of dice and what not. Let's then say have some algorithm that will cluster these objects into things that are similar to each other based on some similarity metric.

Now I haven't told the machine learning algorithm, ahead of time, what categories certain objects belong to. I don't have a cheat sheet that it can learn from where I have a set of existing objects and my correct categorization of it. The machine learning algorithm must infer those categories on its own. This is an example of unsupervised learning, where I don't have a set of answers that I'm getting it learn from. I'm just trying to let the algorithm gather its own answers based on the data presented to it alone.

The problem with this is that we don't necessarily know what the algorithm will come up with! If I gave it that bunch of objects shown in the preceding image, is it going to group things into things that are round, things that are large versus small, things that are red versus blue, I don't know. It's going to depend on the metric that I give it for similarity between items primarily. But sometimes you'll find clusters that are surprising, and emerged that you didn't expect to see.

So that's really the point of unsupervised learning: if you don't know what you're looking for, it can be a powerful tool for discovering classifications that you didn't even know were there. We call this a **latent variable**. Some property of your data that you didn't even know was there originally, can be teased out by unsupervised learning.

Let's take another example around unsupervised learning. Say I was clustering people instead of balls and dice. I'm writing a dating site and I want to see what sorts of people tend to cluster together. There are some attributes that people tend to cluster around, which decide whether they tend to like each other and date each other for example. Now you might find that the clusters that emerge don't conform to your predisposed stereotypes. Maybe it's not about college students versus middle-aged people, or people who are divorced and whatnot, or their religious beliefs. Maybe if you look at the clusters that actually emerged from that analysis, you'll learn something new about your users and actually figure out that there's something more important than any of those existing features of your people that really count toward, to decide whether they like each other. So that's an example of supervised learning providing useful results.

Another example could be clustering movies based on their properties. If you were to run clustering on a set of movies from like IMDb or something, maybe the results would surprise you. Perhaps it's not just about the genre of the movie. Maybe there are other properties, like the age of the movie or the running length or what country it was released in, that are more important. You just never know. Or we could analyze the text of product descriptions and try to find the terms that carry the most meaning for a certain category. Again, we might not necessarily know ahead of time what terms, or what words, are most indicative of a product being in a certain category; but through unsupervised learning, we can tease out that latent information.

Supervised learning

Now in contrast, supervised learning is a case where we have a set of answers that the model can learn from. We give it a set of training data, that the model learns from. It can then infer relationships between the features and the categories that we want, and apply that to unseen new values - and predict information about them.

Going back to our earlier example, where we were trying to predict car prices based on the attributes of those cars. That's an example where we are training our model using actual answers. So I have a set of known cars and their actual prices that they sold for. I train the model on that set of complete answers, and then I can create a model that I'm able to use to predict the prices of new cars that I haven't seen before. That's an example of supervised learning, where you're giving it a set of answers to learn from. You've already assigned categories or some organizing criteria to a set of data, and your algorithm then uses that criteria to build a model from which it can predict new values.

Evaluating supervised learning

So how do you evaluate supervised learning? Well, the beautiful thing about supervised learning is that we can use a trick called train/test. The idea here is to split our observational data that I want my model to learn from into two groups, a training set and a testing set. So when I train/build my model based on the data that I have, I only do that with part of my data that I'm calling my training set, and I reserve another part of my data that I'm going to use for testing purposes.

I can build my model using a subset of my data for training data, and then I'm in a position to evaluate the model that comes out of that, and see if it can successfully predict the correct answers for my testing data.

So you see what I did there? I have a set of data where I already have the answers that I can train my model from, but I'm going to withhold a portion of that data and actually use that to test my model that was generated using the training set! That it gives me a very concrete way to test how good my model is on unseen data because I actually have a bit of data that I set aside that I can test it with.

You can then measure quantitatively how well it did using r-squared or some other metric, like root-mean-square error, for example. You can use that to test one model versus another and see what the best model is for a given problem. You can tune the parameters of that model and use train/test to maximize the accuracy of that model on your testing data. So this is a great way to prevent overfitting.

There are some caveats to supervised learning. need to make sure that both your training and test datasets are large enough to actually be representative of your data. You also need to make sure that you're catching all the different categories and outliers that you care about, in both training and testing, to get a good measure of its success, and to build a good model.

You have to make sure that you've selected from those datasets randomly, and that you're not just carving your dataset in two and saying everything left of here is training and right here is testing. You want to sample that randomly, because there could be some pattern sequentially in your data that you don't know about.

Now, if your model is overfitting, and just going out of its way to accept outliers in your training data, then that's going to be revealed when you put it against unset scene of testing data. This is because all that gyrations for outliers won't help with the outliers that it hasn't seen before.

Let's be clear here that train/test is not perfect, and it is possible to get misleading results from it. Maybe your sample sizes are too small, like we already talked about, or maybe just due to random chance your training data and your test data look remarkably similar, they actually do have a similar set of outliers - and you can still be overfitting. As you can see in the following example, it really can happen:

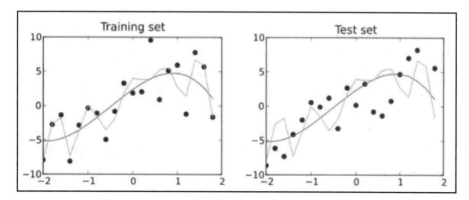

K-fold cross validation

Now there is a way around this problem, called k-fold cross-validation, and we'll look at an example of this later in the book, but the basic concept is you train/test many times. So you actually split your data not into just one training set and one test set, but into multiple randomly assigned segments, k segments. That's where the k comes from. And you reserve one of those segments as your test data, and then you start training your model on the remaining segments and measure their performance against your test dataset. Then you take the average performance from each of those training sets' models' results and take their r-squared average score.

So this way, you're actually training on different slices of your data, measuring them against the same test set, and if you have a model that's overfitting to a particular segment of your training data, then it will get averaged out by the other ones that are contributing to k-fold cross-validation.

Here are the K-fold cross validation steps:

1. Split your data into K randomly-assigned segments
2. Reserve one segment as your test data
3. Train on each of the remaining K-1 segments and measure their performance against the test set
4. Take the average of the K-1 r-squared scores

This will make more sense later in the book, right now I would just like for you to know that this tool exists for actually making train/test even more robust than it already is. So let's go and actually play with some data and actually evaluate it using train/test next.

Using train/test to prevent overfitting of a polynomial regression

Let's put train/test into action. So you might remember that a regression can be thought of as a form of supervised machine learning. Let's just take a polynomial regression, which we covered earlier, and use train/test to try to find the right degree polynomial to fit a given set of data.

Chapter 5

Just like in our previous example, we're going to set up a little fake dataset of randomly generated page speeds and purchase amounts, and I'm going to create a quirky little relationship between them that's exponential in nature.

```
%matplotlib inline
import numpy as np
from pylab import *

np.random.seed(2)

pageSpeeds = np.random.normal(3.0, 1.0, 100)
purchaseAmount = np.random.normal(50.0, 30.0, 100) / pageSpeeds

scatter(pageSpeeds, purchaseAmount)
```

Let's go ahead and generate that data. We'll use a normal distribution of random data for both page speeds and purchase amount using the relationship as shown in the following screenshot:

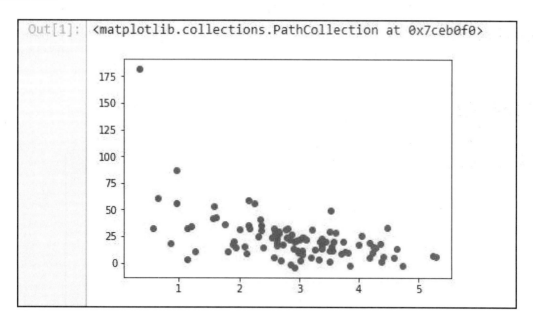

Next, we'll split that data. We'll take 80% of our data, and we're going to reserve that for our training data. So only 80% of these points are going to be used for training the model, and then we're going to reserve the other 20% for testing that model against unseen data.

[165]

We'll use Python's syntax here for splitting the list. The first 80 points are going to go to the training set, and the last 20, everything after 80, is going to go to test set. You may remember this from our Python basics chapter earlier on, where we covered the syntax to do this, and we'll do the same thing for purchase amounts here:

```
trainX = pageSpeeds[:80]
testX = pageSpeeds[80:]

trainY = purchaseAmount[:80]
testY = purchaseAmount[80:]
```

Now in our earlier sections, I've said that you shouldn't just slice your dataset in two like this, but that you should randomly sample it for training and testing. In this case though, it works out because my original data was randomly generated anyway, so there's really no rhyme or reason to where things fell. But in real-world data you'll want to shuffle that data before you split it.

We'll look now at a handy method that you can use for that purpose of shuffling your data. Also, if you're using the pandas package, there's some handy functions in there for making training and test datasets automatically for you. But we're going to do it using a Python list here. So let's visualize our training dataset that we ended up with. We'll do a scatter plot of our training page speeds and purchase amounts.

```
scatter(trainX, trainY)
```

This is what your output should now look like:

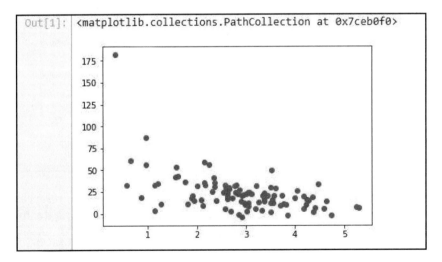

Basically, 80 points that were selected at random from the original complete dataset have been plotted. It has basically the same shape, so that's a good thing. It's representative of our data. That's important!

Now let's plot the remaining 20 points that we reserved as test data.

```
scatter(testX, testY)
```

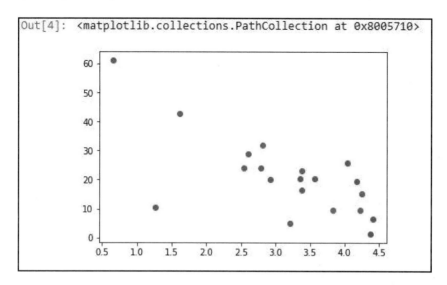

Here, we see our remaining 20 for testing also has the same general shape as our original data. So I think that's a representative test set too. It's a little bit smaller than you would like to see in the real world, for sure. You probably get a little bit of a better result if you had 1,000 points instead of 100, for example, to choose from and reserved 200 instead of 20.

Now we're going to try to fit an 8th degree polynomial to this data, and we'll just pick the number 8 at random because I know it's a really high order and is probably overfitting.

Let's go ahead and fit our 8th degree polynomial using `np.poly1d(np.polyfit(x, y, 8))`, where *x* is an array of the training data only, and *y* is an array of the training data only. We are finding our model using only those 80 points that we reserved for training. Now we have this p4 function that results that we can use to predict new values:

```
x = np.array(trainX)
y = np.array(trainY)

p4 = np.poly1d(np.polyfit(x, y, 8))
```

Now we'll plot the polynomial this came up with against the training data. We can scatter our original data for the training data set, and then we can plot our predicted values against them:

```
import matplotlib.pyplot as plt

xp = np.linspace(0, 7, 100)
axes = plt.axes()
axes.set_xlim([0,7])
axes.set_ylim([0, 200])
plt.scatter(x, y)
plt.plot(xp, p4(xp), c='r')
plt.show()
```

You can see in the following graph that it looks like a pretty good fit, but you know that clearly it's doing some overfitting:

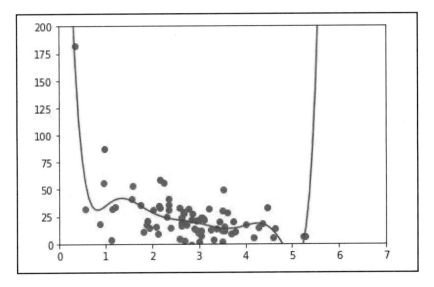

What's this craziness out at the right? I'm pretty sure our real data, if we had it out there, wouldn't be crazy high, as this function would implicate. So this is a great example of overfitting your data. It fits the data you gave it very well, but it would do a terrible job of predicting new values beyond the point where the graph is going crazy high on the right. So let's try to tease that out. Let's give it our test dataset:

```
testx = np.array(testX)
testy = np.array(testY)

axes = plt.axes()
```

```
axes.set_xlim([0,7])
axes.set_ylim([0, 200])
plt.scatter(testx, testy)
plt.plot(xp, p4(xp), c='r')
plt.show()
```

Indeed, if we plot our test data against that same function, well, it doesn't actually look that bad.

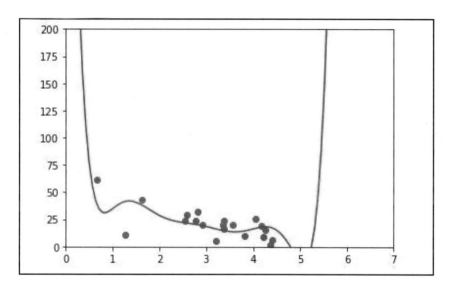

We got lucky and none of our test is actually out here to begin with, but you can see that it's a reasonable fit, but far from perfect. And in fact, if you actually measure the r-squared score, it's worse than you might think. We can measure that using the `r2_score()` function from `sklearn.metrics`. We just give it our original data and our predicted values and it just goes through and measures all the variances from the predictions and squares them all up for you:

```
from sklearn.metrics import r2_score
r2 = r2_score(testy, p4(testx))
print r2
```

We end up with an r-squared score of just 0.3. So that's not that hot! You can see that it fits the training data a lot better:

```
from sklearn.metrics import r2_score
r2 = r2_score(np.array(trainY), p4(np.array(trainX)))
print r2
```

The r-squared value turns out to be 0.6, which isn't too surprising, because we trained it on the training data. The test data is sort of its unknown, its test, and it did fail the test, quite frankly. 30%, that's an F!

So this has been an example where we've used train/test to evaluate a supervised learning algorithm, and like I said before, pandas has some means of making this even easier. We'll look at that a little bit later, and we'll also look at more examples of train/test, including k-fold cross validation, later in the book as well.

Activity

You can probably guess what your homework is. So we know that an 8th order polynomial isn't very useful. Can you do better? So I want you to go back through our example, and use different values for the degree polynomial that you're going to use to fit. Change that 8 to different values and see if you can figure out what degree polynomial actually scores best using train/test as a metric. Where do you get your best r-squared score for your test data? What degree fits here? Go play with that. It should be a pretty easy exercise and a very enlightening one for you as well.

So that's train/test in action, a very important technique to have under your belt, and you're going to use it over and over again to make sure that your results are a good fit for the model that you have, and that your results are a good predictor of unseen values. It's a great way to prevent overfitting when you're doing your modeling.

Bayesian methods - Concepts

Did you ever wonder how the spam classifier in your e-mail works? How does it know that an e-mail might be spam or not? Well, one popular technique is something called Naive Bayes, and that's an example of a Bayesian method. Let's learn more about how that works. Let's discuss Bayesian methods.

We did talk about Bayes' theorem earlier in this book in the context of talking about how things like drug tests could be very misleading in their results. But you can actually apply the same Bayes' theorem to larger problems, like spam classifiers. So let's dive into how that might work, it's called a Bayesian method.

So just a refresher on Bayes' theorem -remember, the probability of A given B is equal to the overall probability of A times the probability of B given A over the overall probability of B:

$$P(A|B) = \frac{P(A)P(B|A)}{P(B)}$$

How can we use that in machine learning? I can actually build a spam classifier for that: an algorithm that can analyze a set of known spam e-mails and a known set of non-spam e-mails, and train a model to actually predict whether new e-mails are spam or not. This is a real technique used in actual spam classifiers in the real world.

As an example, let's just figure out the probability of an e-mail being spam given that it contains the word "free". If people are promising you free stuff, it's probably spam! So let's work that out. The probability of an email being spam given that you have the word "free" in that e-mail works out to the overall probability of it being a spam message times the probability of containing the word "free" given that it's spam over the probability overall of being free:

$$P(Spam \mid Free) = \frac{P(Spam)P(Free \mid Spam)}{P(Free)}$$

The numerator can just be thought of as the probability of a message being Spam and containing the word Free. But that's a little bit different than what we're looking for, because that's the odds out of the complete dataset and not just the odds within things that contain the word Free. The denominator is just the overall probability of containing the word Free. Sometimes that won't be immediately accessible to you from the data that you have. If it's not, you can expand that out to the following expression if you need to derive it:

P(Free|Spam)P(Spam) + P(Free|Not Spam)P(Not Spam))

This gives you the percentage of e-mails that contain the word "free" that are spam, which would be a useful thing to know when you're trying to figure out if it's spam or not.

What about all the other words in the English language, though? So our spam classifier should know about more than just the word "free". It should automatically pick up every word in the message, ideally, and figure out how much does that contribute to the likelihood of a particular e-mail being spam. So what we can do is train our model on every word that we encounter during training, throwing out things like "a" and "the" and "and" and meaningless words like that. Then when we go through all the words in a new e-mail, we can multiply the probability of being spam for each word together, and we get the overall probability of that e-mail being spam.

Now it's called Naive Bayes for a reason. It's naive is because we're assuming that there's no relationships between the words themselves. We're just looking at each word in isolation, individually within a message, and basically combining all the probabilities of each word's contribution to it being spam or not. We're not looking at the relationships between the words. So a better spam classifier would do that, but obviously that's a lot harder.

So this sounds like a lot of work. But the overall idea is not that hard, and scikit-learn in Python makes it actually pretty easy to do. It offers a feature called **CountVectorizer** that makes it very simple to actually split up an e-mail to all of its component words and process those words individually. Then it has a `MultinomialNB` function, where NB stands for Naive Bayes, which will do all the heavy lifting for Naive Bayes for us.

Implementing a spam classifier with Naïve Bayes

Let's write a spam classifier using Naive Bayes. You're going to be surprised how easy this is. In fact, most of the work ends up just being reading all the input data that we're going to train on and actually parsing that data in. The actual spam classification bit, the machine learning bit, is itself just a few lines of code. So that's usually how it works out: reading in and massaging and cleaning up your data is usually most of the work when you're doing data science, so get used to the idea!

```
import os
import io
import numpy
from pandas import DataFrame
from sklearn.feature_extraction.text import CountVectorizer
from sklearn.naive_bayes import MultinomialNB

def readFiles(path):
    for root, dirnames, filenames in os.walk(path):
        for filename in filenames:
            path = os.path.join(root, filename)

            inBody = False
            lines = []
            f = io.open(path, 'r', encoding='latin1')
            for line in f:
                if inBody:
                    lines.append(line)
                elif line == '\n':
                    inBody = True
```

```
            f.close()
            message = '\n'.join(lines)
            yield path, message

def dataFrameFromDirectory(path, classification):
    rows = []
    index = []
    for filename, message in readFiles(path):
        rows.append({'message': message, 'class': classification})
        index.append(filename)

    return DataFrame(rows, index=index)

data = DataFrame({'message': [], 'class': []})

data = data.append(dataFrameFromDirectory(
                  'e:/sundog-consult/Udemy/DataScience/emails/spam',
                  'spam'))
data = data.append(dataFrameFromDirectory(
                  'e:/sundog-consult/Udemy/DataScience/emails/ham',
                  'ham'))
```

So the first thing we need to do is read all those e-mails in somehow, and we're going to again use pandas to make this a little bit easier. Again, pandas is a useful tool for handling tabular data. We import all the different packages that we're going to use within our example here, that includes the os library, the io library, numpy, pandas, and CountVectorizer and MultinomialNB from scikit-learn.

Let's go through this code in detail now. We can skip past the function definitions of `readFiles()` and `dataFrameFromDirectory()` for now and go down to the first thing that our code actually does which is to create a pandas DataFrame object.

We're going to construct this from a dictionary that initially contains a little empty list for messages in an empty list of class. So this syntax is saying, "I want a DataFrame that has two columns: one that contains the message, the actual text of each e-mail; and one that contains the class of each e-mail, that is, whether it's spam or ham". So it's saying I want to create a little database of e-mails, and this database has two columns: the actual text of the e-mail and whether it's spam or not.

Now we needed to put something in that database, that is, into that DataFrame, in Python syntax. So we call the two methods `append()` and `dataFrameFromDirectory()` to actually throw into the DataFrame all the spam e-mails from my spam folder, and all the ham e-mails from the ham folder.

Machine Learning with Python

If you are playing along here, make sure you modify the path passed to the `dataFrameFromDirectory()` function to match wherever you installed the book materials in your system! And again, if you're on Mac or Linux, please pay attention to backslashes and forward slashes and all that stuff. In this case, it doesn't matter, but you won't have a drive letter, if you're not on Windows. So just make sure those paths are actually pointing to where your `spam` and `ham` folders are for this example.

Next, `dataFrameFromDirectory()` is a function I wrote, which basically says I have a path to a directory, and I know it's given classification, spam or ham, then it uses the `readFiles()` function, that I also wrote, which will iterate through every single file in a directory. So `readFiles()` is using the `os.walk()` function to find all the files in a directory. Then it builds up the full pathname for each individual file in that directory, and then it reads it in. And while it's reading it in, it actually skips the header for each e-mail and just goes straight to the text, and it does that by looking for the first blank line.

It knows that everything after the first empty line is actually the message body, and everything in front of that first empty line is just a bunch of header information that I don't actually want to train my spam classifier on. So it gives me back both, the full path to each file and the body of the message. So that's how we read in all of the data, and that's the majority of the code!

So what I have at the end of the day is a DataFrame object, basically a database with two columns, that contains message bodies, and whether it's spam or not. We can go ahead and run that, and we can use the `head` command from the DataFrame to actually preview what this looks like:

```
data.head()
```

The first few entries in our DataFrame look like this: for each path to a given file full of e-mails we have a classification and we have the message body:

	class	message
C:\Users\deveshc\Desktop\DataScience\emails\spam\00001.7848dde101aa985090474a91ec93fcf0	spam	<!DOCTYPE HTML PUBLIC "-//W3C//DTD H'
C:\Users\deveshc\Desktop\DataScience\emails\spam\00002.d94f1b97e48ed3b553b3508d116e6a09	spam	1) Fight The Risk of Cancer!\n\nhttp://www.ad
C:\Users\deveshc\Desktop\DataScience\emails\spam\00003.2ee33bc6eacdb11f38d052c44819ba6c	spam	1) Fight The Risk of Cancer!\n\nhttp://www.ad
C:\Users\deveshc\Desktop\DataScience\emails\spam\00004.eac8de8d759b7e74154f142194282724	spam	##
C:\Users\deveshc\Desktop\DataScience\emails\spam\00005.57696a39d7d84318ce497886896bf90d	spam	I thought you might like these:\n\n1) Slim Dov

Alright, now for the fun part, we're going to use the `MultinomialNB()` function from scikit-learn to actually perform Naive Bayes on the data that we have.

```
vectorizer = CountVectorizer()
counts = vectorizer.fit_transform(data['message'].values)

classifier = MultinomialNB()
targets = data['class'].values
classifier.fit(counts, targets)
```

This is what your output should now look like:

```
Out[13]: MultinomialNB(alpha=1.0, class_prior=None, fit_prior=True)
```

Once we build a `MultinomialNB` classifier, it needs two inputs. It needs the actual data that we're training on (`counts`), and the targets for each thing (`targets`). So `counts` is basically a list of all the words in each e-mail and the number of times that word occurs.

So this is what `CountVectorizer()` does: it takes the `message` column from the DataFrame and takes all the values from it. I'm going to call `vectorizer.fit_transform` which basically tokenizes or converts all the individual words seen in my data into numbers, into values. It then counts up how many times each word occurs.

This is a more compact way of representing how many times each word occurs in an e-mail. Instead of actually preserving the words themselves, I'm representing those words as different values in a sparse matrix, which is basically saying that I'm treating each word as a number, as a numerical index, into an array. What that does is, just in plain English, it split each message up into a list of words that are in it, and counts how many times each word occurs. So we're calling that `counts`. It's basically that information of how many times each word occurs in each individual message. Mean while `targets` is the actual classification data for each e-mail that I've encountered. So I can call `classifier.fit()` using my `MultinomialNB()` function to actually create a model using Naive Bayes, which will predict whether new e-mails are spam or not based on the information we've given it.

Machine Learning with Python

Let's go ahead and run that. It runs pretty quickly! I'm going to use a couple of examples here. Let's try a message body that just says `Free Money now!!!` which is pretty clearly spam, and a more innocent message that just says `"Hi Bob, how about a game of golf tomorrow?"` So we're going to pass these in.

```
examples = ['Free Money now!!!', "Hi Bob, how about a game of golf
tomorrow?"]
example_counts = vectorizer.transform(examples)
predictions = classifier.predict(example_counts)
predictions
```

The first thing we do is convert the messages into the same format that I trained my model on. So I use that same vectorizer that I created when creating the model to convert each message into a list of words and their frequencies, where the words are represented by positions in an array. Then once I've done that transformation, I can actually use the `predict()` function on my classifier, on that array of examples that have transformed into lists of words, and see what we come up with:

```
Out[14]: array(['spam', 'ham'],
         dtype='|S4')
```

`array(['spam', 'ham'], dtype='|S4')`

And sure enough, it works! So, given this array of two input messages, `Free Money now!!!` and `Hi Bob`, it's telling me that the first result came back as spam and the second result came back as ham, which is what I would expect. That's pretty cool. So there you have it.

Activity

We had a pretty small dataset here, so you could try running some different e-mails through it if you want and see if you get different results. If you really want to challenge yourself, try applying train/test to this example. So the real measure of whether or not my spam classifier is good or not is not just intuitively whether it can figure out that `Free Money now!!!` is spam. You want to measure that quantitatively.

So if you want a little bit of a challenge, go ahead and try to split this data up into a training set and a test dataset. You can actually look up online how pandas can split data up into train sets and testing sets pretty easily for you, or you can do it by hand. Whatever works for you. See if you can actually apply your `MultinomialNB` classifier to a test dataset and measure its performance. So, if you want a little bit of an exercise, a little bit of a challenge, go ahead and give that a try.

How cool is that? We just wrote our own spam classifier just using a few lines of code in Python. It's pretty easy using scikit-learn and Python. That's Naive Bayes in action, and you can actually go and classify some spam or ham messages now that you have that under your belt. Pretty cool stuff. Let's talk about clustering next.

K-Means clustering

Next, we're going to talk about k-means clustering, and this is an unsupervised learning technique where you have a collection of stuff that you want to group together into various clusters. Maybe it's movie genres or demographics of people, who knows? But it's actually a pretty simple idea, so let's see how it works.

K-means clustering is a very common technique in machine learning where you just try to take a bunch of data and find interesting clusters of things just based on the attributes of the data itself. Sounds fancy, but it's actually pretty simple. All we do in k-means clustering is try to split our data into K groups - that's where the K comes from, it's how many different groups you're trying to split your data into - and it does this by finding K centroids.

So, basically, what group a given data point belongs to is defined by which of these centroid points it's closest to in your scatter plot. You can visualize this in the following image:

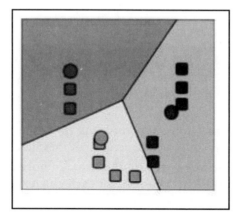

This is showing an example of k-means clustering with K of three, and the squares represent data points in a scatter plot. The circles represent the centroids that the k-means clustering algorithm came up with, and each point is assigned a cluster based on which centroid it's closest to. So that's all there is to it, really. It's an example of unsupervised learning. It isn't a case where we have a bunch of data and we already know the correct cluster for a given set of training data; rather, you're just given the data itself and it tries to converge on these clusters naturally just based on the attributes of the data alone. It's also an example where you are trying to find clusters or categorizations that you didn't even know were there. As with most unsupervised learning techniques, the point is to find latent values, things you didn't really realize were there until the algorithm showed them to you.

For example, where do millionaires live? I don't know, maybe there is some interesting geographical cluster where rich people tend to live, and k-means clustering could help you figure that out. Maybe I don't really know if today's genres of music are meaningful. What does it mean to be alternative these days? Not much, right? But by using k-means clustering on attributes of songs, maybe I could find interesting clusters of songs that are related to each other and come up with new names for what those clusters represent. Or maybe I can look at demographic data, and maybe existing stereotypes are no longer useful. Maybe Hispanic has lost its meaning and there's actually other attributes that define groups of people, for example, that I could uncover with clustering. Sounds fancy, doesn't it? Really complicated stuff. Unsupervised machine learning with K clusters, it sounds fancy, but as with most techniques in data science, it's actually a very simple idea.

Here's the algorithm for us in plain English:

1. **Randomly pick K centroids (k-means):** We start off with a randomly chosen set of centroids. So if we have a K of three we're going to look for three clusters in our group, and we will assign three randomly positioned centroids in our scatter plot.
2. **Assign each data point to the centroid it is closest to:** We then assign each data point to the randomly assigned centroid that it is closest to.
3. **Recompute the centroids based on the average position of each centroid's points**: Then recompute the centroid for each cluster that we come up with. That is, for a given cluster that we end up with, we will move that centroid to be the actual center of all those points.
4. **Iterate until points stop changing assignment to centroids:** We will do it all again until those centroids stop moving, we hit some threshold value that says OK, we have converged on something here.
5. **Predict the cluster for new points:** To predict the clusters for new points that I haven't seen before, we can just go through our centroid locations and figure out which centroid it's closest to to predict its cluster.

Let's look at a graphical example to make a little bit more sense. We'll call the first figure in the following image as A, second as B, third as C and the fourth as D.

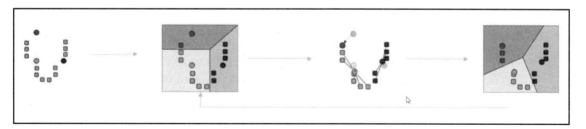

The gray squares in image A represent data points in our scatter plot. The axes represent some different features of something. Maybe it's age and income; it's an example I keep using, but it could be anything. And the gray squares might represent individual people or individual songs or individual something that I want to find relationships between.

So I start off by just picking three points at random on my scatterplot. Could be anywhere. Got to start somewhere, right? The three points (centroids) I selected have been shown as circles in image A. So the next thing I'm going to do is for each centroid I'll compute which one of the gray points it's closest to. By doing that, the points shaded in blue are associated with this blue centroid. The green points are closest to the green centroid, and this single red point is closest to that red random point that I picked out.

Of course, you can see that's not really reflective of where the actual clusters appear to be. So what I'm going to do is take the points that ended up in each cluster and compute the actual center of those points. For example, in the green cluster, the actual center of all data turns out to be a little bit lower. We're going to move the centroid down a little bit. The red cluster only had one point, so its center moves down to where that single point is. And the blue point was actually pretty close to the center, so that just moves a little bit. On this next iteration we end up with something that looks like image D. Now you can see that our cluster for red things has grown a little bit and things have moved a little bit, that is, those got taken from the green cluster.

If we do that again, you can probably predict what's going to happen next. The green centroid will move a little bit, the blue centroid will still be about where it is. But at the end of the day you're going to end up with the clusters you'd probably expect to see. That's how k-means works. So it just keeps iterating, trying to find the right centroids until things start moving around and we converge on a solution.

Limitations to k-means clustering

So there are some limitations to k-means clustering. Here they are:

1. **Choosing K:** First of all, we need to choose the right value of K, and that's not a straightforward thing to do at all. The principal way of choosing K is to just start low and keep increasing the value of K depending on how many groups you want, until you stop getting large reductions in squared error. If you look at the distances from each point to their centroids, you can think of that as an error metric. At the point where you stop reducing that error metric, you know you probably have too many clusters. So you're not really gaining any more information by adding additional clusters at that point.

2. **Avoiding local minima:** Also, there is a problem of local minima. You could just get very unlucky with those initial choices of centroids and they might end up just converging on local phenomena instead of more global clusters, so usually, you want to run this a few times and maybe average the results together. We call that ensemble learning. We'll talk about that more a little bit later on, but it's always a good idea to run k-means more than once using a different set of random initial values and just see if you do in fact end up with the same overall results or not.

3. **Labeling the clusters:** Finally, the main problem with k-means clustering is that there's no labels for the clusters that you get. It will just tell you that this group of data points are somehow related, but you can't put a name on it. It can't tell you the actual meaning of that cluster. Let's say I have a bunch of movies that I'm looking at, and k-means clustering tells me that bunch of science fiction movies are over here, but it's not going to call them "science fiction" movies for me. It's up to me to actually dig into the data and figure out, well, what do these things really have in common? How might I describe that in English? That's the hard part, and k-means won't help you with that. So again, scikit-learn makes it very easy to do this.

Let's now work up an example and put k-means clustering into action.

Clustering people based on income and age

Let's see just how easy it is to do k-means clustering using scikit-learn and Python.

The first thing we're going to do is create some random data that we want to try to cluster. Just to make it easier, we'll actually build some clusters into our fake test data. So let's pretend there's some real fundamental relationship between these data, and there are some real natural clusters that exist in it.

So to do that, we can work with this little `createClusteredData()` function in Python:

```
from numpy import random, array

#Create fake income/age clusters for N people in k clusters
def createClusteredData(N, k):
    random.seed(10)
    pointsPerCluster = float(N)/k
    X = []
    for i in range (k):
        incomeCentroid = random.uniform(20000.0, 200000.0)
        ageCentroid = random.uniform(20.0, 70.0)
        for j in range(int(pointsPerCluster)):
            X.append([random.normal(incomeCentroid, 10000.0),
                random.normal(ageCentroid, 2.0)])
    X = array(X)
    return X
```

The function starts off with a consistent random seed so you'll get the same result every time. We want to create clusters of N people in k clusters. So we pass `N` and `k` to `createClusteredData()`.

Our code figures out how many points per cluster that works out to first and stores it in `pointsPerCluster`. Then, it builds up list `X` that starts off empty. For each cluster, we're going to create some random centroid of income (`incomeCentroid`) between 20,000 and 200,000 dollars and some random centroid of age (`ageCentroid`) between the age of 20 and 70.

Machine Learning with Python

What we're doing here is creating a fake scatter plot that will show income versus age for N people and k clusters. So for each random centroid that we created, I'm then going to create a normally distributed set of random data with a standard deviation of 10,000 in income and a standard deviation of 2 in age. That will give us back a bunch of age income data that is clustered into some pre-existing clusters that we can chose at random. OK, let's go ahead and run that.

Now, to actually do k-means, you'll see how easy it is.

```
from sklearn.cluster import KMeans
import matplotlib.pyplot as plt
from sklearn.preprocessing import scale
from numpy import random, float

data = createClusteredData(100, 5)

model = KMeans(n_clusters=5)

# Note I'm scaling the data to normalize it! Important for good results.
model = model.fit(scale(data))

# We can look at the clusters each data point was assigned to
print model.labels_

# And we'll visualize it:
plt.figure(figsize=(8, 6))
plt.scatter(data[:,0], data[:,1], c=model.labels_.astype(float))
plt.show()
```

All you need to do is import `KMeans` from scikit-learn's `cluster` package. We're also going to import `matplotlib` so we can visualize things, and also import `scale` so we can take a look at how that works.

So we use our `createClusteredData()` function to say 100 random people around 5 clusters. So there are 5 natural clusters for the data that I'm creating. We then create a model, a KMeans model with k of 5, so we're picking 5 clusters because we know that's the right answer. But again, in unsupervised learning you don't necessarily know what the real value of k is. You need to iterate and converge on it yourself. And then we just call `model.fit` using my KMeans `model` using the data that we had.

Now the scale I alluded to earlier, that's normalizing the data. One important thing with k-means is that it works best if your data is all normalized. That means everything is at the same scale. So a problem that I have here is that my ages range from 20 to 70, but my incomes range all the way up to 200,000. So these values are not really comparable. The incomes are much larger than the age values. `Scale` will take all that data and scale it together to a consistent scale so I can actually compare these things as apples to apples, and that will help a lot with your k-means results.

So, once we've actually called `fit` on our model, we can actually look at the resulting labels that we got. Then we can actually visualize it using a little bit of `matplotlib` magic. You can see in the code we have a little trick where we assigned the color to the labels that we ended up with converted to some floating point number. That's just a little trick you can use to assign arbitrary colors to a given value. So let's see what we end up with:

```
[0 0 0 0 0 0 0 0 0 0 0 0 0 0 0 0 0 0 0 0 1 1 1 1 1 1 1 1 1 1 1 1 1 1 1 1
 1 1 1 4 4 4 4 4 4 4 4 4 4 4 4 4 4 4 4 4 4 4 4 4 2 3 3 3 3 3 3 3 3 3 3 3 3 3
 3 3 3 3 3 3 2 2 2 2 2 2 2 2 2 2 2 2 2 2 2 2 2 2 2 2]
```

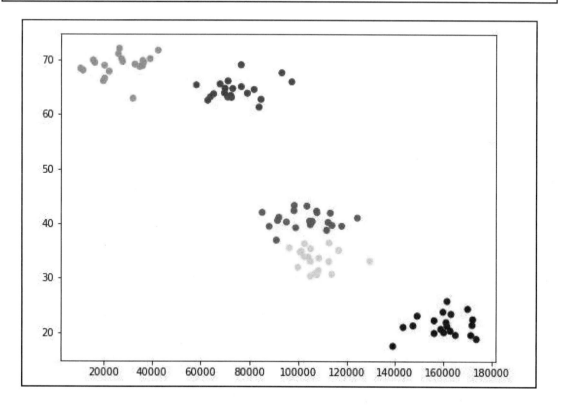

It didn't take that long. You see the results are basically what clusters I assigned everything into. We know that our fake data is already pre-clustered, so it seems that it identified the first and second clusters pretty easily. It got a little bit confused beyond that point, though, because our clusters in the middle are actually a little bit mushed together. They're not really that distinct, so that was a challenge for k-means. But regardless, it did come up with some reasonable guesses at the clusters. This is probably an example of where four clusters would more naturally fit the data.

Activity

So what I want you to do for an activity is to try a different value of k and see what you end up with. Just eyeballing the preceding graph, it looks like four would work well. Does it really? What happens if I increase k too large? What happens to my results? What does it try to split things into, and does it even make sense? So, play around with it, try different values of `k`. So in the `n_clusters()` function, change the 5 to something else. Run all through it again and see you end up with.

That's all there is to k-means clustering. It's just that simple. You can just use scikit-learn's `KMeans` thing from `cluster`. The only real gotcha: make sure you scale the data, normalize it. You want to make sure the things that you're using k-means on are comparable to each other, and the `scale()` function will do that for you. So those are the main things for k-means clustering. Pretty simple concept, even simpler to do it using scikit-learn.

That's all there is to it. That's k-means clustering. So if you have a bunch of data that is unclassified and you don't really have the right answers ahead of time, it's a good way to try to naturally find interesting groupings of your data, and maybe that can give you some insight into what that data is. It's a good tool to have. I've used it before in the real world and it's really not that hard to use, so keep that in your tool chest.

Measuring entropy

Quite soon we're going to get to one of the cooler parts of machine learning, at least I think so, called decision trees. But before we can talk about that, it's a necessary to understand the concept of entropy in data science.

So entropy, just like it is in physics and thermodynamics, is a measure of a dataset's disorder, of how same or different the dataset is. So imagine we have a dataset of different classifications, for example, animals. Let's say I have a bunch of animals that I have classified by species. Now, if all of the animals in my dataset are an iguana, I have very low entropy because they're all the same. But if every animal in my dataset is a different animal, I have iguanas and pigs and sloths and who knows what else, then I would have a higher entropy because there's more disorder in my dataset. Things are more different than they are the same.

Entropy is just a way of quantifying that sameness or difference throughout my data. So, an entropy of 0 implies all the classes in the data are the same, whereas if everything is different, I would have a high entropy, and something in between would be a number in between. Entropy just describes how same or different the things in a dataset are.

Now mathematically, it's a little bit more involved than that, so when I actually compute a number for entropy, it's computed using the following expression:

$$H(S) = -p_1 \ln p_1 - \cdots - p_n \ln p_n$$

p_i represents the proportion of the data labeled for each class

So for every different class that I have in my data, I'm going to have one of these p terms, p_1, p_2, and so on and so forth through p_n, for n different classes that I might have. The p just represents the proportion of the data that is that class. And if you actually plot what this looks like for each term- `pi* ln * pi`, it'll look a little bit something like the following graph:

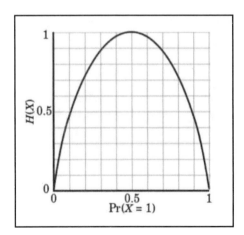

You add these up for each individual class. For example, if the proportion of the data, that is, for a given class is 0, then the contribution to the overall entropy is 0. And if everything is that class, then again the contribution to the overall entropy is 0 because in either case, if nothing is this class or everything is this class, that's not really contributing anything to the overall entropy.

It's the things in the middle that contribute entropy of the class, where there's some mixture of this classification and other stuff. When you add all these terms together, you end up with an overall entropy for the entire dataset. So mathematically, that's how it works out, but again, the concept is very simple. It's just a measure of how disordered your dataset, how same or different the things in your data are.

Decision trees - Concepts

Believe it or not, given a set of training data, you can actually get Python to generate a flowchart for you to make a decision. So if you have something you're trying to predict on some classification, you can use a decision tree to actually look at multiple attributes that you can decide upon at each level in the flowchart. You can print out an actual flowchart for you to use to make a decision from, based on actual machine learning. How cool is that? Let's see how it works.

I personally find decision trees are one of the most interesting applications of machine learning. A decision tree basically gives you a flowchart of how to make some decision. You have some dependent variable, like whether or not I should go play outside today or not based on the weather. When you have a decision like that that depends on multiple attributes or multiple variables, a decision tree could be a good choice.

There are many different aspects of the weather that might influence my decision of whether I should go outside and play. It might have to do with the humidity, the temperature, whether it's sunny or not, for example. A decision tree can look at all these different attributes of the weather, or anything else, and decide what are the thresholds? What are the decisions I need to make on each one of those attributes before I arrive at a decision of whether or not I should go play outside? That's all a decision tree is. So it's a form of supervised learning.

The way it would work in this example would be as follows. I would have some sort of dataset of historical weather, and data about whether or not people went outside to play on a particular day. I would feed the model this data of whether it was sunny or not on each day, what the humidity was, and if it was windy or not; and whether or not it was a good day to go play outside. Given that training data, a decision tree algorithm can then arrive at a tree that gives us a flowchart that we can print out. It looks just like the following flow chart. You can just walk through and figure out whether or not it's a good day to play outside based on the current attributes. You can use that to predict the decision for a new set of values:

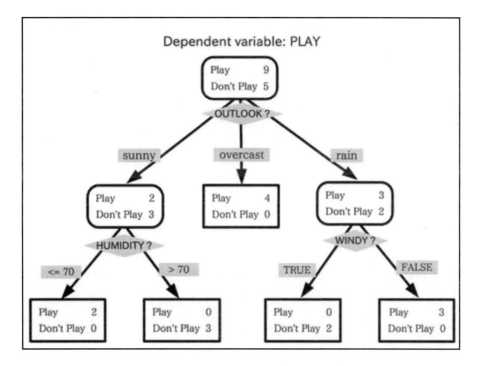

How cool is that? We have an algorithm that will make a flowchart for you automatically just based on observational data. What's even cooler is how simple it all works once you learn how it works.

Decision tree example

Let's say I want to build a system that will automatically filter out resumes based on the information in them. A big problem that technology companies have is that we get tons and tons of resumes for our positions. We have to decide who we actually bring in for an interview, because it can be expensive to fly somebody out and actually take the time out of the day to conduct an interview. So what if there were a way to actually take historical data on who actually got hired and map that to things that are found on their resume?

We could construct a decision tree that will let us go through an individual resume and say, "OK, this person actually has a high likelihood of getting hired, or not". We can train a decision tree on that historical data and walk through that for future candidates. Wouldn't that be a wonderful thing to have?

So let's make some totally fabricated hiring data that we're going to use in this example:

Candidate ID	Years Experience	Employed?	Previous employers	Level of Education	Top-tier school	Interned	Hired
0	10	1	4	0	0	0	1
1	0	0	0	0	1	1	1
2	7	0	6	0	0	0	0
3	2	1	1	1	1	0	1
4	20	0	2	2	1	0	0

In the preceding table, we have candidates that are just identified by numerical identifiers. I'm going to pick some attributes that I think might be interesting or helpful to predict whether or not they're a good hire or not. How many years of experience do they have? Are they currently employed? How many employers have they had previous to this one? What's their level of education? What degree do they have? Did they go to what we classify as a top-tier school? Did they do an internship while they were in college? We can take a look at this historical data, and the dependent variable here is `Hired`. Did this person actually get a job offer or not based on that information?

Now, obviously there's a lot of information that isn't in this model that might be very important, but the decision tree that we train from this data might actually be useful in doing an initial pass at weeding out some candidates. What we end up with might be a tree that looks like the following:

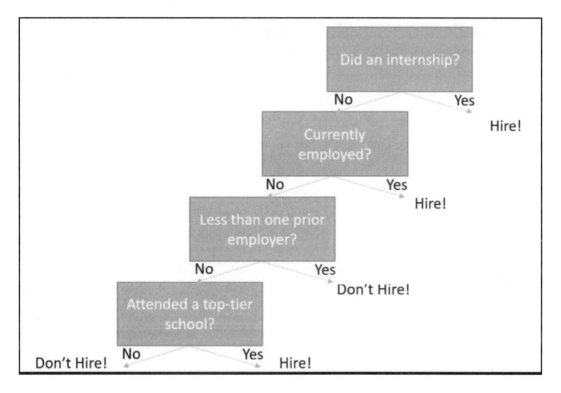

- So it just turns out that in my totally fabricated data, anyone that did an internship in college actually ended up getting a job offer. So my first decision point is "did this person do an internship or not?" If yes, go ahead and bring them in. In my experience, internships are actually a pretty good predictor of how good a person is. If they have the initiative to actually go out and do an internship, and actually learn something at that internship, that's a good sign.
- Do they currently have a job? Well, if they are currently employed, in my very small fake dataset it turned out that they are worth hiring, just because somebody else thought they were worth hiring too. Obviously it would be a little bit more of a nuanced decision in the real world.
- If they're not currently employed, do they have less than one prior employer? If yes, this person has never held a job and they never did an internship either. Probably not a good hire decision. Don't hire that person.
- But if they did have a previous employer, did they at least go to a top-tier school? If not, it's kind of iffy. If so, then yes, we should hire this person based on the data that we trained on.

Walking through a decision tree

So that's how you walk through the results of a decision tree. It's just like going through a flowchart, and it's kind of awesome that an algorithm can produce this for you. The algorithm itself is actually very simple. Let me explain how the algorithm works.

At each step of the decision tree flowchart, we find the attribute that we can partition our data on that minimizes the entropy of the data at the next step. So we have a resulting set of classifications: in this case hire or don't hire, and we want to choose the attribute decision at that step that will minimize the entropy at the next step.

At each step we want to make all of the remaining choices result in either as many no hires or as many hire decisions as possible. We want to make that data more and more uniform so as we work our way down the flowchart, and we ultimately end up with a set of candidates that are either all hires or all no hires so we can classify into yes/no decisions on a decision tree. So we just walk down the tree, minimize entropy at each step by choosing the right attribute to decide on, and we keep on going until we run out.

There's a fancy name for this algorithm. It's called **ID3 (Iterative Dichotomiser 3)**. It is what's known as a greedy algorithm. So as it goes down the tree, it just picks the attribute that will minimize entropy at that point. Now that might not actually result in an optimal tree that minimizes the number of choices that you have to make, but it will result in a tree that works, given the data that you gave it.

Random forests technique

Now one problem with decision trees is that they are very prone to overfitting, so you can end up with a decision tree that works beautifully for the data that you trained it on, but it might not be that great for actually predicting the correct classification for new people that it hasn't seen before. Decision trees are all about arriving at the right decision for the training data that you gave it, but maybe you didn't really take into account the right attributes, maybe you didn't give it enough of a representative sample of people to learn from. This can result in real problems.

So to combat this issue, we use a technique called random forests, where the idea is that we sample the data that we train on, in different ways, for multiple different decision trees. Each decision tree takes a different random sample from our set of training data and constructs a tree from it. Then each resulting tree can vote on the right result.

Now that technique of randomly resampling our data with the same model is a term called bootstrap aggregating, or bagging. This is a form of what we call ensemble learning, which we'll cover in more detail shortly. But the basic idea is that we have multiple trees, a forest of trees if you will, each that uses a random subsample of the data that we have to train on. Then each of these trees can vote on the final result, and that will help us combat overfitting for a given set of training data.

The other thing random forests can do is actually restrict the number of attributes that it can choose, between at each stage, while it is trying to minimize the entropy as it goes. And we can randomly pick which attributes it can choose from at each level. So that also gives us more variation from tree to tree, and therefore we get more of a variety of algorithms that can compete with each other. They can all vote on the final result using slightly different approaches to arriving at the same answer.

So that's how random forests work. Basically, it is a forest of decision trees where they are drawing from different samples and also different sets of attributes at each stage that it can choose between.

So, with all that, let's go make some decision trees. We'll use random forests as well when we're done, because scikit-learn makes it really really easy to do, as you'll see soon.

Decision trees - Predicting hiring decisions using Python

Turns out that it's easy to make decision trees; in fact it's crazy just how easy it is, with just a few lines of Python code. So let's give it a try.

I've included a `PastHires.csv` file with your book materials, and that just includes some fabricated data, that I made up, about people that either got a job offer or not based on the attributes of those candidates.

```
import numpy as np
import pandas as pd
from sklearn import tree

input_file = "c:/spark/DataScience/PastHires.csv"
df = pd.read_csv(input_file, header = 0)
```

Machine Learning with Python

You'll want to please immediately change that path I used here for my own system (`c:/spark/DataScience/PastHires.csv`) to wherever you have installed the materials for this book. I'm not sure where you put it, but it's almost certainly not there.

We will use `pandas` to read our CSV in, and create a DataFrame object out of it. Let's go ahead and run our code, and we can use the `head()` function on the DataFrame to print out the first few lines and make sure that it looks like it makes sense.

```
df.head()
```

Sure enough we have some valid data in the output:

Out[2]:

	Years Experience	Employed?	Previous employers	Level of Education	Top-tier school	Interned	Hired
0	10	Y	4	BS	N	N	Y
1	0	N	0	BS	Y	Y	Y
2	7	N	6	BS	N	N	N
3	2	Y	1	MS	Y	N	Y
4	20	N	2	PhD	Y	N	N

So, for each candidate ID, we have their years of past experience, whether or not they were employed, their number of previous employers, their highest level of education, whether they went to a top-tier school, and whether they did an internship; and finally here, in the Hired column, the answer - where we knew that we either extended a job offer to this person or not.

As usual, most of the work is just in massaging your data, preparing your data, before you actually run the algorithms on it, and that's what we need to do here. Now scikit-learn requires everything to be numerical, so we can't have Ys and Ns and BSs and MSs and PhDs. We have to convert all those things to numbers for the decision tree model to work. The way to do this is to use some short-hand in pandas, which makes these things easy. For example:

```
d = {'Y': 1, 'N': 0}
df['Hired'] = df['Hired'].map(d)
df['Employed?'] = df['Employed?'].map(d)
df['Top-tier school'] = df['Top-tier school'].map(d)
df['Interned'] = df['Interned'].map(d)
d = {'BS': 0, 'MS': 1, 'PhD': 2}
df['Level of Education'] = df['Level of Education'].map(d)
df.head()
```

Basically, we're making a dictionary in Python that maps the letter Y to the number 1, and the letter N to the value 0. So, we want to convert all our Ys to 1s and Ns to 0s. So 1 will mean yes and 0 will mean no. What we do is just take the Hired column from the DataFrame, and call `map()` on it, using a dictionary. This will go through the entire Hired column, in the entire DataFrame and use that dictionary lookup to transform all the entries in that column. It returns a new DataFrame column that I'm putting back into the Hired column. This replaces the Hired column with one that's been mapped to 1s and 0s.

We do the same thing for Employed, Top-tier school and Interned, so all those get mapped using the yes/no dictionary. So, the Ys and Ns become 1s and 0s instead. For the Level of Education, we do the same trick, we just create a dictionary that assigns BS to 0, MS to 1, and PhD to 2 and uses that to remap those degree names to actual numerical values. So if I go ahead and run that and do a `head()` again, you can see that it worked:

Out[3]:

	Years Experience	Employed?	Previous employers	Level of Education	Top-tier school	Interned	Hired
0	10	1	4	0	0	0	1
1	0	0	0	0	1	1	1
2	7	0	6	0	0	0	0
3	2	1	1	1	1	0	1
4	20	0	2	2	1	0	0

All my yeses are 1's, my nos are 0's, and my Level of Education is now represented by a numerical value that has real meaning.

Next we need to prepare everything to actually go into our decision tree classifier, which isn't that hard. To do that, we need to separate our feature information, which are the attributes that we're trying to predict from, and our target column, which contains the thing that we're trying to predict. To extract the list of feature name columns, we are just going to create a list of columns up to number 6. We go ahead and print that out.

```
features = list(df.columns[:6])
features
```

We get the following output:

```
Out[4]: ['Years Experience',
         'Employed?',
         'Previous employers',
         'Level of Education',
         'Top-tier school',
         'Interned']
```

Above are the column names that contain our feature information: Years Experience, Employed?, Previous employers, Level of Education, Top-tier school, and Interned. These are the attributes of candidates that we want to predict hiring on.

Next, we construct our *y* vector which is assigned what we're trying to predict, that is our Hired column:

```
y = df["Hired"]
X = df[features]
clf = tree.DecisionTreeClassifier()
clf = clf.fit(X,y)
```

This code extracts the entire Hired column and calls it y. Then it takes all of our columns for the feature data and puts them in something called X. This is a collection of all of the data and all of the feature columns, and X and y are the two things that our decision tree classifier needs.

To actually create the classifier itself, two lines of code: we call `tree.DecisionTreeClassifier()` to create our classifier, and then we fit it to our feature data (X) and the answers (y)- whether or not people were hired. So, let's go ahead and run that.

Displaying graphical data is a little bit tricky, and I don't want to distract us too much with the details here, so please just consider the following boilerplate code. You don't need to get into how Graph viz works here - and dot files and all that stuff: it's not important to our journey right now. The code you need to actually display the end results of a decision tree is simply:

```
from IPython.display import Image
from sklearn.externals.six import StringIO
import pydot

dot_data = StringIO()
tree.export_graphviz(clf, out_file=dot_data,
                     feature_names=features)
```

Chapter 5

```
graph = pydot.graph_from_dot_data(dot_data.getvalue())
Image(graph.create_png())
```

So let's go ahead and run this.

This is what your output should now look like:

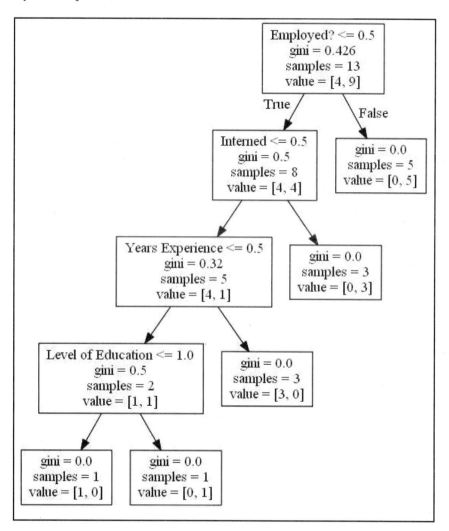

There we have it! How cool is that?! We have an actual flow chart here.

[195]

Now, let me show you how to read it. At each stage, we have a decision. Remember most of our data which is yes or no, is going to be **0** or **1**. So, the first decision point becomes: is Employed? less than **0.5**? Meaning that if we have an employment value of 0, that is no, we're going to go left.If employment is 1, that is yes, we're going to go right.

So, were they previously employed? If not go left, if yes go right. It turns out that in my sample data, everyone who is currently employed actually got a job offer, so I can very quickly say if you are currently employed, yes, you're worth bringing in, we're going to follow down to the second level here.

So, how do you interpret this? The gini score is basically a measure of entropy that it's using at each step. Remember as we're going down the algorithm is trying to minimize the amount of entropy. And the samples are the remaining number of samples that haven't beensectioned off by a previous decision.

So say this person was employed. The way to read the right leaf node is the value column that tells you at this point we have 0 candidates that were no hires and 5 that were hires. So again, the way to interpret the first decision point is if Employed? was 1, I'm going to go to the right, meaning that they are currently employed, and this brings me to a world where everybody got a job offer. So, that means I should hire this person.

Now let's say that this person doesn't currently have a job. The next thing I'm going to look at is, do they have an internship. If yes, then we're at a point where in our training data everybody got a job offer. So, at that point, we can say our entropy is now 0 (`gini=0.0000`), because everyone's the same, and they all got an offer at that point. However, you know if we keep going down(where the person has not done an internship),we'll be at a point where the entropy is 0.32. It's getting lower and lower, that's a good thing.

Next we're going to look at how much experience they have, do they have less than one year of experience? And, if the case is that they do have some experience and they've gotten this far they're a pretty good no hire decision. We end up at the point where we have zero entropy but, all three remaining samples in our training set were no hires. We have 3 no hires and 0 hires. But, if they do have less experience, then they're probably fresh out of college, they still might be worth looking at.

The final thing we're going to look at is whether or not they went to a Top-tier school, and if so, they end up being a good prediction for being a hire. If not, they end up being a no hire. We end up with one candidate that fell into that category that was a no hire and 0 that were a hire. Whereas, in the case candidates did go to a top tier school, we have 0 no hires and 1 hire.

So, you can see we just keep going until we reach an entropy of 0, if at all possible, for every case.

Ensemble learning – Using a random forest

Now, let's say we want to use a random forest, you know, we're worried that we might be over fitting our training data. It's actually very easy to create a random forest classifier of multiple decision trees.

So, to do that, we can use the same data that we created before. You just need your *X* and *y* vectors, that is the set of features and the column that you're trying to predict on:

```
from sklearn.ensemble import RandomForestClassifier

clf = RandomForestClassifier(n_estimators=10)
clf = clf.fit(X, y)

#Predict employment of an employed 10-year veteran
print clf.predict([[10, 1, 4, 0, 0, 0]])
#...and an unemployed 10-year veteran
print clf.predict([[10, 0, 4, 0, 0, 0]])
```

We make a random forest classifier, also available from scikit-learn, and pass it the number of trees we want in our forest. So, we made ten trees in our random forest in the code above. We then fit that to the model.

You don't have to walk through the trees by hand, and when you're dealing with a random forest you can't really do that anyway. So, instead we use the `predict()` function on the model, that is on the classifier that we made. We pass in a list of all the different features for a given candidate that we want to predict employment for.

If you remember this maps to these columns: Years Experience, Employed?, Previous employers, Level of Education, Top-tier school, and Interned; interpreted as numerical values. We predict the employment of an employed 10-year veteran. We also predict the employment of an unemployed 10-year veteran. And, sure enough, we get a result:

```
[1]
[1]
```

So, in this particular case, we ended up with a hire decision on both. But, what's interesting is there is a random component to that. You don't actually get the same result every time! More often than not, the unemployed person does not get a job offer, and if you keep running this you'll see that's usually the case. But, the random nature of bagging, of bootstrap aggregating each one of those trees, means you're not going to get the same result every time. So, maybe 10 isn't quite enough trees. So, anyway, that's a good lesson to learn here!

Activity

For an activity, if you want to go back and play with this, mess around with my input data. Go ahead and edit the code we've been exploring, and create an alternate universe where it's a topsy turvy world; for example, everyone that I gave a job offer to now doesn't get one and vice versa. See what that does to your decision tree. Just mess around with it and see what you can do and try to interpret the results.

So, that's decision trees and random forests, one of the more interesting bits of machine learning, in my opinion. I always think it's pretty cool to just generate a flowchart out of thin air like that. So, hopefully you'll find that useful.

Ensemble learning

When we talked about random forests, that was an example of ensemble learning, where we're actually combining multiple models together to come up with a better result than any single model could come up with. So, let's learn about that in a little bit more depth. Let's talk about ensemble learning a little bit more.

So, remember random forests? We had a bunch of decision trees that were using different subsamples of the input data, and different sets of attributes that it would branch on, and they all voted on the final result when you were trying to classify something at the end. That's an example of ensemble learning. Another example: when we were talking about k-means clustering, we had the idea of maybe using different k-means models with different initial random centroids, and letting them all vote on the final result as well. That is also an example of ensemble learning.

Basically, the idea is that you have more than one model, and they might be the same kind of model or it might be different kinds of models, but you run them all, on your set of training data, and they all vote on the final result for whatever it is you're trying to predict. And oftentimes, you'll find that this ensemble of different models produces better results than any single model could on its own.

A good example, from a few years ago, was the Netflix prize. Netflix ran a contest where they offered, I think it was a million dollars, to any researcher who could outperform their existing movie recommendation algorithm. The ones that won were ensemble approaches, where they actually ran multiple recommender algorithms at once and let them all vote on the final result. So, ensemble learning can be a very powerful, yet simple tool, for increasing the quality of your final results in machine learning. Let us now try to explore various types of ensemble learning:

- **Bootstrap aggregating or bagging:** Now, random forests use a technique called bagging, short for bootstrap aggregating. This means that we take random subsamples of our training data and feed them into different versions of the same model and let them all vote on the final result. If you remember, random forests took many different decision trees that use a different random sample of the training data to train on, and then they all came together in the end to vote on a final result. That's bagging.
- **Boosting:** Boosting is an alternate model, and the idea here is that you start with a model, but each subsequent model boosts the attributes that address the areas that were misclassified by the previous model. So, you run train/tests on a model, you figure out what are the attributes that it's basically getting wrong, and then you boost those attributes in subsequent models - in hopes that those subsequent models will pay more attention to them, and get them right. So, that's the general idea behind boosting. You run a model, figure out its weak points, amplify the focus on those weak points as you go, and keep building more and more models that keep refining that model, based on the weaknesses of the previous one.
- **Bucket of models:** Another technique, and this is what that Netflix prize-winner did, is called a bucket of models, where you might have entirely different models that try to predict something. Maybe I'm using k-means, a decision tree, and regression. I can run all three of those models together on a set of training data and let them all vote on the final classification result when I'm trying to predict something. And maybe that would be better than using any one of those models in isolation.

- **Stacking:** Stacking has the same idea. So, you run multiple models on the data, combine the results together somehow. The subtle difference here between bucket of models and stacking, is that you pick the model that wins. So, you'd run train/test, you find the model that works best for your data, and you use that model. By contrast, stacking will combine the results of all those models together, to arrive at a final result.

Now, there is a whole field of research on ensemble learning that tries to find the optimal ways of doing ensemble learning, and if you want to sound smart, usually that involves using the word Bayes a lot. So, there are some very advanced methods of doing ensemble learning but all of them have weak points, and I think this is yet another lesson in that we should always use the simplest technique that works well for us.

Now these are all very complicated techniques that I can't really get into in the scope of this book, but at the end of the day, it's hard to outperform just the simple techniques that we've already talked about. A few of the complex techniques are listed here:

- **Bayes optical classifier:** In theory, there's something called the Bayes Optimal Classifier that will always be the best, but it's impractical, because it's computationally prohibitive to do it.
- **Bayesian parameter averaging:** Many people have tried to do variations of the Bayes Optimal Classifier to make it more practical, like the Bayesian Parameter Averaging variation. But it's still susceptible to overfitting and it's often outperformed by bagging, which is the same idea behind random forests; you just resample the data multiple times, run different models, and let them all vote on the final result. Turns out that works just as well, and it's a heck of a lot simpler!
- **Bayesian model combination:** Finally, there's something called Bayesian Model Combination that tries to solve all the shortcomings of Bayes Optimal Classifier and Bayesian Parameter Averaging. But, at the end of the day, it doesn't do much better than just cross validating against the combination of models.

Again, these are very complex techniques that are very difficult to use. In practice, we're better off with the simpler ones that we've talked about in more detail. But, if you want to sound smart and use the word Bayes a lot it's good to be familiar with these techniques at least, and know what they are.

So, that's ensemble learning. Again, the takeaway is that the simple techniques, like bootstrap aggregating, or bagging, or boosting, or stacking, or bucket of models, are usually the right choices. There are some much fancier techniques out there but they're largely theoretical. But, at least you know about them now.

It's always a good idea to try ensemble learning out. It's been proven time and time again that it will produce better results than any single model, so definitely consider it!

Support vector machine overview

Finally, we're going to talk about **support vector machines** (**SVM**), which is a very advanced way of clustering or classifying higher dimensional data.

So, what if you have multiple features that you want to predict from? SVM can be a very powerful tool for doing that, and the results can be scarily good! It's very complicated under the hood, but the important things are understanding when to use it, and how it works at a higher level. So, let's cover SVM now.

Support vector machines is a fancy name for what actually is a fancy concept. But fortunately, it's pretty easy to use. The important thing is knowing what it does, and what it's good for. So, support vector machines works well for classifying higher-dimensional data, and by that I mean lots of different features. So, it's easy to use something like k-means clustering, to cluster data that has two dimensions, you know, maybe age on one axis and income on another. But, what if I have many, many different features that I'm trying to predict from. Well, support vector machines might be a good way of doing that.

Support vector machines finds higher-dimensional support vectors across which to divide the data (mathematically, these support vectors define hyperplanes). That is, mathematically, what support vector machines can do is find higher dimensional support vectors (that's where it gets its name from) that define the higher-dimensional planes that split the data into different clusters.

Obviously the math gets pretty weird pretty quickly with all this. Fortunately, the `scikit-learn` package will do it all for you, without you having to actually get into it. Under the hood, you need to understand though that it uses something called the kernel trick to actually find those support vectors or hyperplanes that might not be apparent in lower dimensions. There are different kernels you can use, to do this in different ways. The main point is that SVM's are a good choice if you have higher- dimensional data with lots of different features, and there are different kernels you can use that have varying computational costs and might be better fits for the problem at hand.

The important point is that SVMs employ some advanced mathematical trickery to cluster data, and it can handle data sets with lots of features. It's also fairly expensive - the "kernel trick" is the only thing that makes it possible.

Machine Learning with Python

I want to point out that SVM is a supervised learning technique. So, we're actually going to train it on a set of training data, and we can use that to make predictions for future unseen data or test data. It's a little bit different than k-means clustering and that k-means was completely unsupervised; with a support vector machine, by contrast, it is training based on actual training data where you have the answer of the correct classification for some set of data that it can learn from. So, SVM's are useful for classification and clustering, if you will - but it's a supervised technique!

One example that you often see with SVMs is using something called support vector classification. The typical example uses the Iris dataset which is one of the sample datasets that comes with scikit-learn. This set is a classification of different flowers, different observations of different Iris flowers and their species. The idea is to classify these using information about the length and width of the petal on each flower, and the length and width of the sepal of each flower. (The sepal, apparently, is a little support structure underneath the petal. I didn't know that until now either.) You have four dimensions of attributes there; you have the length and width of the petal, and the length and the width of the sepal. You can use that to predict the species of an Iris given that information.

Here's an example of doing that with SVC: basically, we have sepal width and sepal length projected down to two dimensions so we can actually visualize it:

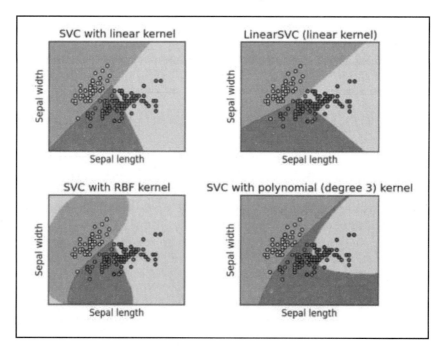

With different kernels you might get different results. SVC with a linear kernel will produce something very much as you see in the preceding image. You can use polynomial kernels or fancier kernels that might project down to curves in two dimensions as shown in the image. You can do some pretty fancy classification this way.

These have increasing computational costs, and they can produce more complex relationships. But again, it's a case where too much complexity can yield misleading results, so you need to be careful and actually use train/test when appropriate. Since we are doing supervised learning, you can actually do train/test and find the right model that works, or maybe use an ensemble approach.

You need to arrive at the right kernel for the task at hand. For things like polynomial SVC, what's the right degree polynomial to use? Even things like linear SVC will have different parameters associated with them that you might need to optimize for. This will make more sense with a real example, so let's dive into some actual Python code and see how it works!

Using SVM to cluster people by using scikit-learn

Let's try out some support vector machines here. Fortunately, it's a lot easier to use than it is to understand. We're going to go back to the same example I used for k-means clustering, where I'm going to create some fabricated cluster data about ages and incomes of a hundred random people.

If you want to go back to the k-means clustering section, you can learn more about kind of the idea behind this code that generates the fake data. And if you're ready, please consider the following code:

```
import numpy as np

#Create fake income/age clusters for N people in k clusters
def createClusteredData(N, k):
    pointsPerCluster = float(N)/k
    X = []
    y = []
    for i in range (k):
        incomeCentroid = np.random.uniform(20000.0, 200000.0)
        ageCentroid = np.random.uniform(20.0, 70.0)
        for j in range(int(pointsPerCluster)):
            X.append([np.random.normal(incomeCentroid, 10000.0),
                np.random.normal(ageCentroid, 2.0)])
            y.append(i)
```

```
X = np.array(X)
y = np.array(y)
return X, y
```

Please note that because we're using supervised learning here, we not only need the feature data again, but we also need the actual answers for our training dataset.

What the `createClusteredData()` function does here, is to create a bunch of random data for people that are clustered around `k` points, based on age and income, and it returns two arrays. The first array is the feature array, that we're calling X, and then we have the array of the thing we're trying to predict for, which we're calling y. A lot of times in scikit-learn when you're creating a model that you can predict from, those are the two inputs that it will take, a list of feature vectors, and the thing that you're trying to predict, that it can learn from. So, we'll go ahead and run that.

So now we're going to use the `createClusteredData()` function to create 100 random people with 5 different clusters. We will just create a scatter plot to illustrate those, and see where they land up:

```
%matplotlib inline
from pylab import *

(X, y) = createClusteredData(100, 5)

plt.figure(figsize=(8, 6))
plt.scatter(X[:,0], X[:,1], c=y.astype(np.float))
plt.show()
```

The following graph shows our data that we're playing with. Every time you run this you're going to get a different set of clusters. So, you know, I didn't actually use a random seed... to make life interesting.

A couple of new things here--I'm using the `figsize` parameter on `plt.figure()` to actually make a larger plot. So, if you ever need to adjust the size in matplotlib, that's how you do it. I'm using that same trick of using the color as the classification number that I end up with. So the number of the cluster that I started with is being plotted as the color of these data points. You can see, it's a pretty challenging problem, there's definitely some intermingling of clusters here:

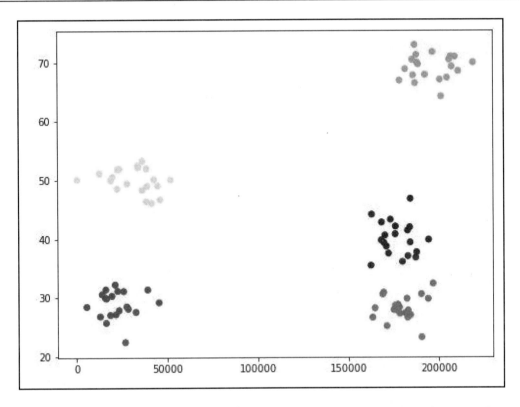

Now we can use linear SVC (SVC is a form of SVM), to actually partition that into clusters. We're going to use SVM with a linear kernel, and with a C value of 1.0. C is just an error penalty term that you can adjust; it's 1 by default. Normally, you won't want to mess with that, but if you're doing some sort of convergence on the right model using ensemble learning or train/test, that's one of the things you can play with. Then, we will fit that model to our feature data, and the actual classifications that we have for our training dataset.

```
from sklearn import svm, datasets

C = 1.0
svc = svm.SVC(kernel='linear', C=C).fit(X, y)
```

So, let's go ahead and run that. I don't want to get too much into how we're actually going to visualize the results here, just take it on faith that `plotPredictions()` is a function that can plot the classification ranges and SVC.

Machine Learning with Python

It helps us visualize where different classifications come out. Basically, it's creating a mesh across the entire grid, and it will plot different classifications from the SVC models as different colors on that grid, and then we're going to plot our original data on top of that:

```
def plotPredictions(clf):
    xx, yy = np.meshgrid(np.arange(0, 250000, 10),
                         np.arange(10, 70, 0.5))
    Z = clf.predict(np.c_[xx.ravel(), yy.ravel()])

    plt.figure(figsize=(8, 6))
    Z = Z.reshape(xx.shape)
    plt.contourf(xx, yy, Z, cmap=plt.cm.Paired, alpha=0.8)
    plt.scatter(X[:,0], X[:,1], c=y.astype(np.float))
    plt.show()

plotPredictions(svc)
```

So, let's see how that works out. SVC is computationally expensive, so it takes a long time to run:

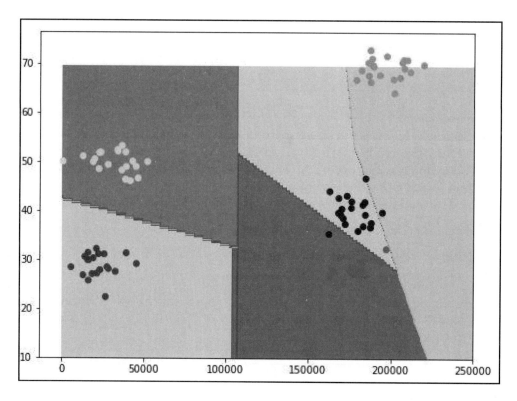

You can see here that it did its best. Given that it had to draw straight lines, and polygonal shapes, it did a decent job of fitting to the data that we had. So, you know, it did miss a few - but by and large, the results are pretty good.

SVC is actually a very powerful technique; it's real strength is in higher dimensional feature data. Go ahead and play with it. By the way if you want to not just visualize the results, you can use the `predict()` function on the SVC model, just like on pretty much any model in scikit-learn, to pass in a feature array that you're interested in. If I want to predict the classification for someone making $200,000 a year who was 40 years old, I would use the following code:

```
svc.predict([[200000, 40]])
```

This would put that person in, in our case, cluster number 1:

```
Out[5]:  array([3])
```

If I had a someone making $50,000 here who was 65, I would use the following code:

```
svc.predict([[50000, 65]])
```

This is what your output should now look like:

```
Out[6]:  array([4])
```

That person would end up in cluster number 2, whatever that represents in this example. So, go ahead and play with it.

Activity

Now, linear is just one of many kernels that you can use, like I said there are many different kernels you can use. One of them is a polynomial model, so you might want to play with that. Please do go ahead and look up the documentation. It's good practice for you to looking at the docs. If you're going to be using scikit-learn in any sort of depth, there's a lot of different capabilities and options that you have available to you. So, go look up scikit-learn online, find out what the other kernels are for the SVC method, and try them out, see if you actually get better results or not.

This is a little exercise, not just in playing with SVM and different kinds of SVC, but also in familiarizing yourself with how to learn more on your own about SVC. And, honestly, a very important trait of any data scientist or engineer is going to be the ability to go and look up information yourself when you don't know the answers.

So, you know, I'm not being lazy by not telling you what those other kernels are, I want you to get used to the idea of having to look this stuff up on your own, because if you have to ask someone else about these things all the time you're going to get really annoying, really fast in a workplace. So, go look that up, play around it, see what you come up with.

So, that's SVM/SVC, a very high power technique that you can use for classifying data, in supervised learning. Now you know how it works and how to use it, so keep that in your bag of tricks!

Summary

In this chapter, we saw some interesting machine learning techniques. We covered one of the fundamental concepts behind machine learning called train/test. We saw how to use train/test to try to find the right degree polynomial to fit a given set of data. We then analyzed the difference between supervised and unsupervised machine learning.

We saw how to implement a spam classifier and enable it to determine whether an email is spam or not using the Naive Bayes technique. We talked about k-means clustering, an unsupervised learning technique, which helps group data into clusters. We also looked at an example using scikit-learn which clustered people based on their income and age.

We then went on to look at the concept of entropy and how to measure it. We walked through the concept of decision trees and how, given a set of training data, you can actually get Python to generate a flowchart for you to actually make a decision. We also built a system that automatically filters out resumes based on the information in them and predicts the hiring decision of a person.

We learned along the way the concept of ensemble learning, and we concluded by talking about support vector machines, which is a very advanced way of clustering or classifying higher dimensional data. We then moved on to use SVM to cluster people using scikit-learn. In the next chapter, we'll talk about recommender systems.

6
Recommender Systems

Let's talk about my personal area of expertise—recommender systems, so systems that can recommend stuff to people based on what everybody else did. We'll look at some examples of this and a couple of ways to do it. Specifically, two techniques called user-based and item-based collaborative filtering. So, let's dive in.

I spent most of my career at `amazon.com` and `imdb.com`, and a lot of what I did there was developing recommender systems; things like *people who bought this also bought*, or *recommended for you*, and things that did movie recommendations for people. So, this is something I know a lot about personally, and I hope to share some of that knowledge with you. We'll walk through, step by step, covering the following topics:

- What are recommender systems?
- User-based collaborative filtering
- Item-based collaborative filtering
- Finding movie similarities
- Making movie recommendations to people
- Improving the recommender's results

What are recommender systems?

Well, like I said Amazon is a great example, and one I'm very familiar with. So, if you go to their recommendations section, as shown in the following image, you can see that it will recommend things that you might be interested in purchasing based on your past behavior on the site.

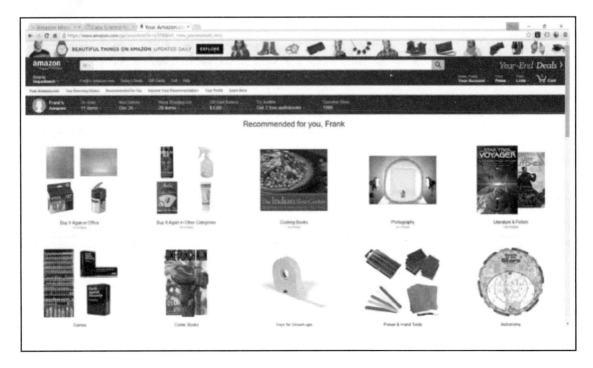

The recommender system might include things that you've rated, or things that you bought, and other data as well. I can't go into the details because they'll hunt me down, and you know, do bad things to me. But, it's pretty cool. You can also think of the *people who bought this also bought* feature on Amazon as a form of recommender system.

The difference is that the recommendations you're seeing on your Amazon recommendations page are based on all of your past behavior, whereas people *who bought this also bought* or *people who viewed this also viewed*, things like that, are just based on the thing you're looking at right now, and showing you things that are similar to it that you might also be interested in. And, it turns out, what you're doing right now is probably the strongest signal of your interest anyhow.

Another example is from Netflix, as shown in the following image (the following image is a screenshot from Netflix):

They have various features that try to recommend new movies or other movies you haven't seen yet, based on the movies that you liked or watched in the past as well, and they break that down by genre. They have kind of a different spin on things, where they try to identify the genres or the types of movies that they think you're enjoying the most and they then show you more results from those genres. So, that's another example of a recommender system in action.

The whole point of it is to help you discover things you might not know about before, so it's pretty cool. You know, it gives individual movies, or books, or music, or whatever, a chance to be discovered by people who might not have heard about them before. So, you know, not only is it cool technology, it also kind of levels the playing field a little bit, and helps new items get discovered by the masses. So, it plays a very important role in today's society, at least I'd like to think so! There are few ways of doing this, and we'll look at the main ones in this chapter.

User-based collaborative filtering

First, let's talk about recommending stuff based on your past behavior. One technique is called user-based collaborative filtering, and here's how it works:

> Collaborative filtering, by the way, is just a fancy name for saying recommending stuff based on the combination of what you did and what everybody else did, okay? So, it's looking at your behavior and comparing that to everyone else's behavior, to arrive at the things that might be interesting to you that you haven't heard of yet.

1. The idea here is we build up a matrix of everything that every user has ever bought, or viewed, or rated, or whatever signal of interest that you want to base the system on. So basically, we end up with a row for every user in our system, and that row contains all the things they did that might indicate some sort of interest in a given product. So, picture a table, I have users for the rows, and each column is an item, okay? That might be a movie, a product, a web page, whatever; you can use this for many different things.
2. I then use that matrix to compute the similarity between different users. So, I basically treat each row of this as a vector and I can compute the similarity between each vector of users, based on their behavior.
3. Two users who liked mostly the same things would be very similar to each other and I can then sort this by those similarity scores. If I can find all the users similar to you based on their past behavior, I can then find the users most similar to me, and recommend stuff that they liked that I didn't look at yet.

Let's look at a real example, and it'll make a little bit more sense:

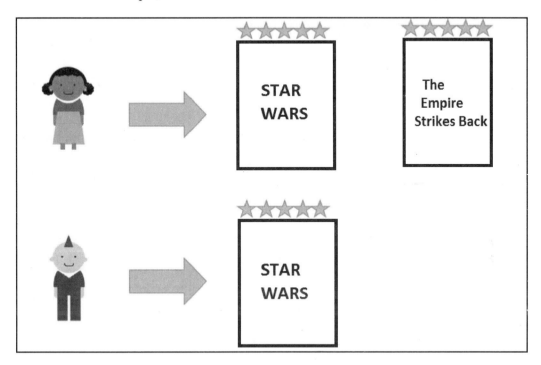

Let's say that this nice lady in the preceding image watched Star Wars and The Empire Strikes Back and she loved them both. So, we have a user vector, of this lady, giving a 5-star rating to Star Wars and The Empire Strikes Back.

Let's also say Mr. Edgy Mohawk Man comes along and he only watched Star Wars. That's the only thing he's seen, he doesn't know about The Empire Strikes Back yet, somehow, he lives in some strange universe where he doesn't know that there are actually many, many Star Wars movies, growing every year in fact.

We can of course say that this guy's actually similar to this other lady because they both enjoyed Star Wars a lot, so their similarity score is probably fairly good and we can say, okay, well, what has this lady enjoyed that he hasn't seen yet? And, The Empire Strikes Back is one, so we can then take that information that these two users are similar based on their enjoyment of Star Wars, find that this lady also liked The Empire Strikes Back, and then present that as a good recommendation for Mr. Edgy Mohawk Man.

We can then go ahead and recommend The Empire Strikes Back to him and he'll probably love it, because in my opinion, it's actually a better film! But I'm not going to get into geek wars with you here.

Limitations of user-based collaborative filtering

Now, unfortunately, user-based collaborative filtering has some limitations. When we think about relationships and recommending things based on relationships between items and people and whatnot, our mind tends to go on relationships between people. So, we want to find people that are similar to you and recommend stuff that they liked. That's kind of the intuitive thing to do, but it's not the best thing to do! The following is the list of some limitations of user-based collaborative filtering:

- One problem is that people are fickle; their tastes are always changing. So, maybe that nice lady in the previous example had sort of a brief science fiction action film phase that she went through and then she got over it, and maybe later in her life she started getting more into dramas or romance films or romcoms. So, what would happen if my Edgy Mohawk guy ended up with a high similarity to her just based on her earlier sci-fi period, and we ended up recommending romantic comedies to him as a result? That would be bad. I mean, there is some protection against that in terms of how we compute the similarity scores to begin with, but it still pollutes our data that people's tastes can change over time. So, comparing people to people isn't always a straightforward thing to do, because people change.
- The other problem is that there's usually a lot more people than there are things in your system, so 7 billion people in the world and counting, there's probably not 7 billion movies in the world, or 7 billion items that you might be recommending out of your catalog. The computational problem finding all the similarities between all of the users in your system is probably much greater than the problem of finding similarities between the items in your system. So, by focusing the system on users, you're making your computational problem a lot harder than it might need to be, because you have a lot of users, at least hopefully you do if you're working for a successful company.
- The final problem is that people do bad things. There's a very real economic incentive to make sure that your product or your movie or whatever it is gets recommended to people, and there are people who try to game the system to make that happen for their new movie, or their new product, or their new book, or whatever.

> It's pretty easy to fabricate fake personas in the system by creating a new user and having them do a sequence of events that likes a lot of popular items and then likes your item too. This is called a **shilling attack**, and we want to ideally have a system that can deal with that.

[214]

There is research around how to detect and avoid these shilling attacks in user-based collaborative filtering, but an even better approach would be to use a totally different approach entirely that's not so susceptible to gaming the system.

That's user-based collaborative filtering. Again, it's a simple concept-you look at similarities between users based on their behavior, and recommend stuff that a user enjoyed that was similar to you, that you haven't seen yet. Now, that does have its limitations as we talked about. So, let's talk about flipping the whole thing on its head, with a technique called item-based collaborative filtering.

Item-based collaborative filtering

Let's now try to address some of the shortcomings in user-based collaborative filtering with a technique called item-based collaborative filtering, and we'll see how that can be more powerful. It's actually one of the techniques that Amazon uses under the hood, and they've talked about this publicly so I can tell you that much, but let's see why it's such a great idea. With user-based collaborative filtering we base our recommendations on relationships between people, but what if we flip that and base them on relationships between items? That's what item-based collaborative filtering is.

Understanding item-based collaborative filtering

This is going to draw on a few insights. For one thing, we talked about people being fickle-their tastes can change over time, so comparing one person to another person based on their past behavior becomes pretty complicated. People have different phases where they have different interests, and you might not be comparing the people that are in the same phase to each other. But, an item will always be whatever it is. A movie will always be a movie, it's never going to change. Star Wars will always be Star Wars, well until George Lucas tinkers with it a little bit, but for the most part, items do not change as much as people do. So, we know that these relationships are more permanent, and there's more of a direct comparison you can make when computing similarity between items, because they do not change over time.

The other advantage is that there are generally fewer things that you're trying to recommend than there are people you're recommending to. So again, 7 billion people in the world, you're probably not offering 7 billion things on your website to recommend to them, so you can save a lot of computational resources by evaluating relationships between items instead of users, because you will probably have fewer items than you have users in your system. That means you can run your recommendations more frequently, make them more current, more up-to-date, and better! You can use more complicated algorithms because you have less relationships to compute, and that's a good thing!

It's also harder to game the system. So, we talked about how easy it is to game a user-based collaborative filtering approach by just creating some fake users that like a bunch of popular stuff and then the thing you're trying to promote. With item-based collaborative filtering that becomes much more difficult. You have to game the system into thinking there are relationships between items, and since you probably don't have the capability to create fake items with fake ties to other items based on many, many other users, it's a lot harder to game an item-based collaborative filtering system, which is a good thing.

While I'm on the topic of gaming the system, another important thing is to make sure that people are voting with their money. A general technique for avoiding shilling attacks or people trying to game your recommender system, is to make sure that the signal behavior is based on people actually spending money. So, you're always going to get better and more reliable results when you base recommendations on what people actually bought, as opposed to what they viewed or what they clicked on, okay?

How item-based collaborative filtering works?

Alright, let's talk about how item-based collaborative filtering works. It's very similar to user-based collaborative filtering, but instead of users, we're looking at items.

So, let's go back to the example of movie recommendations. The first thing we would do is find every pair of movies that is watched by the same person. So, we go through and find every movie that was watched by identical people, and then we measure the similarity of all those people who viewed that movie to each other. So, by this means we can compute similarities between two different movies, based on the ratings of the people who watched both of those movies.

So, let's presume I have a movie pair, okay? Maybe Star Wars and The Empire Strikes Back. I find a list of everyone who watched both of those movies, then I compare their ratings to each other, and if they're similar then I can say these two movies are similar, because they were rated similarly by people who watched both of them. That's the general idea here. That's one way to do it, there's more than one way to do it!

And then I can just sort everything by the movie, and then by the similarity strength of all the similar movies to it, and there's my results for *people who liked also liked*, or *people who rated this highly also rated this highly* and so on and so forth. And like I said, that's just one way of doing it.

That's step one of item-based collaborative filtering-first I find relationships between movies based on the relationships of the people who watched every given pair of movies. It'll make more sense when we go through the following example:

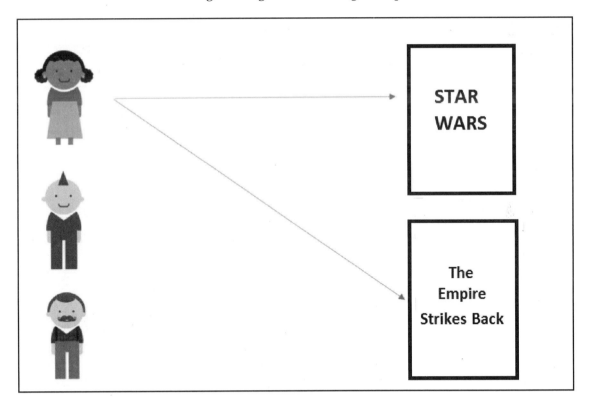

For example, let's say that our nice young lady in the preceding image watched Star Wars and The Empire Strikes Back and liked both of them, so rated them both five stars or something. Now, along comes Mr. Edgy Mohawk Man who also watched Star Wars and The Empire Strikes Back and also liked both of them. So, at this point we can say there's a relationship, there is a similarity between Star Wars and The Empire Strikes Back based on these two users who liked both movies.

What we're going to do is look at each pair of movies. We have a pair of Star Wars and Empire Strikes Back, and then we look at all the users that watched both of them, which are these two guys, and if they both liked them, then we can say that they're similar to each other. Or, if they both disliked them we can also say they're similar to each other, right? So, we're just looking at the similarity score of these two users' behavior related to these two movies in this movie pair.

So, along comes Mr. Moustachy Lumberjack Hipster Man and he watches The Empire Strikes Back and he lives in some strange world where he watched The Empire Strikes Back, but had no idea that Star Wars the first movie existed.

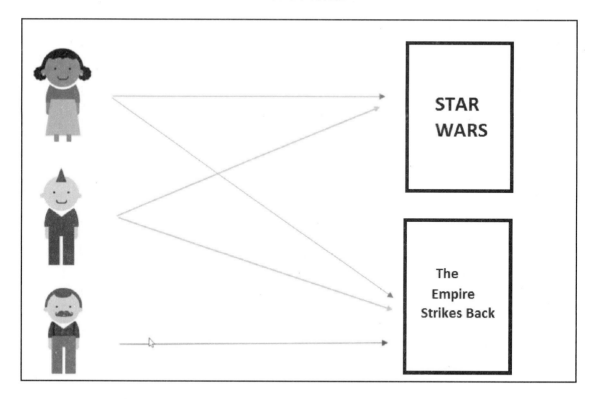

Well that's fine, we computed a relationship between The Empire Strikes Back and Star Wars based on the behavior of these two people, so we know that these two movies are similar to each other. So, given that Mr. Hipster Man liked The Empire Strikes Back, we can say with good confidence that he would also like Star Wars, and we can then recommend that back to him as his top movie recommendation. Something like the following illustration:

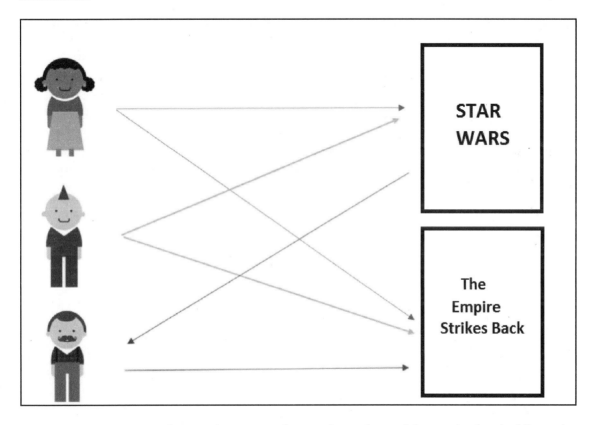

You can see that you end up with very similar results in the end, but we've kind of flipped the whole thing on its head. So, instead of focusing the system on relationships between people, we're focusing them on relationships between items, and those relationships are still based on the aggregate behavior of all the people that watch them. But fundamentally, we're looking at relationships between items and not relationships between people. Got it?

Collaborative filtering using Python

Alright, so let's do it! We have some Python code that will use Pandas, and all the various other tools at our disposal, to create movie recommendations with a surprisingly little amount of code.

The first thing we're going to do is show you item-based collaborative filtering in practice. So, we'll build up *people who watched also watched* basically, you know, *people who rated things highly also rated this thing highly*, so building up these movie to movie relationships. So, we're going to base it on real data that we got from the MovieLens project. So, if you go to MovieLens.org, there's actually an open movie recommender system there, where people can rate movies and get recommendations for new movies.

And, they make all the underlying data publicly available for researchers like us. So, we're going to use some real movie ratings data-it is a little bit dated, it's like 10 years old, so keep that in mind, but it is real behavior data that we're going to be working with finally here. And, we will use that to compute similarities between movies. And, that data in and of itself is useful. You can use that data to say *people who liked also liked*. So, let's say I'm looking at a web page for a movie. the system can then say: *if you liked this movie, and given that you're looking at it you're probably interested in it, then you might also like these movies*. And that's a form of a recommender system right there, even though we don't even know who you are.

Now, it is real-world data, so we're going to encounter some real-world problems with it. Our initial set of results aren't going to look good, so we're going to spend a little bit of extra time trying to figure out why, which is a lot of what you spend your time doing as a data scientist-correct those problems, and go back and run it again until we get results that makes sense.

And finally, we'll actually do item-based collaborative filtering in its entirety, where we actually recommend movies to individuals based on their own behavior. So, let's do this, let's get started!

Finding movie similarities

Let's apply the concept of item-based collaborative filtering. To start with, movie similarities-figure out what movies are similar to other movies. In particular, we'll try to figure out what movies are similar to Star Wars, based on user rating data, and we'll see what we get out of it. Let's dive in!

Okay so, let's go ahead and compute the first half of item-based collaborative filtering, which is finding similarities between items. Download and open the `SimilarMovies.ipynb` file.

Finding Similar Movies

We'll start by loading up the MovieLens dataset. Using Pandas, we can very quickly load the rows of the u.data and u.item files that we care about, and merge them together so we can work with movie names instead of ID's. (In a real production job, you'd stick with ID's and worry about the names at the display layer to make things more efficient. But this lets us understand what's going on better for now.)

In [1]:
```
import pandas as pd

r_cols = ['user_id', 'movie_id', 'rating']
ratings = pd.read_csv('e:/sundog-consult/udemy/datascience/ml-100k/u.data', sep='\t', names=r_cols, usecols=range(3))

m_cols = ['movie_id', 'title']
movies = pd.read_csv('e:/sundog-consult/udemy/datascience/ml-100k/u.item', sep='|', names=m_cols, usecols=range(2))

ratings = pd.merge(movies, ratings)
```

In [2]: `ratings.head()`

Out[2]:

	movie_id	title	user_id	rating
0	1	Toy Story (1995)	308	4
1	1	Toy Story (1995)	287	5
2	1	Toy Story (1995)	148	4
3	1	Toy Story (1995)	280	4
4	1	Toy Story (1995)	66	3

In this case, we're going to be looking at similarities between movies, based on user behavior. And, we're going to be using some real movie rating data from the GroupLens project. GroupLens.org provides real movie ratings data, by real people who are using the `MovieLens.org` website to rate movies and get recommendations back for new movies that they want to watch.

We have included the data files that you need from the GroupLens dataset with the course materials, and the first thing we need to do is import those into a Pandas DataFrame, and we're really going to see the full power of Pandas in this example. It's pretty cool stuff!

Understanding the code

The first thing we're going to do is import the u.data file as part of the MovieLens dataset, and that is a tab-delimited file that contains every rating in the dataset.

```
import pandas as pd

r_cols = ['user_id', 'movie_id', 'rating']
ratings = pd.read_csv('e:/sundog-consult/packt/datascience/ml-100k/u.data',
                sep='\\t', names=r_cols, usecols=range(3))
```

Note that you'll need to add the path here to where you stored the downloaded MovieLens files on your computer. So, the way that this works is even though we're calling read_csv on Pandas, we can specify a different separator than a comma. In this case, it's a tab.

We're basically saying take the first three columns in the u.data file, and import it into a new DataFrame, with three columns: user_id, movie_id, and rating.

What we end up with here is a DataFrame that has a row for every user_id, which identifies some person, and then, for every movie they rated, we have the movie_id, which is some numerical shorthand for a given movie, so Star Wars might be movie 53 or something, and their rating, you know, 1 to 5 stars. So, we have here a database, a DataFrame, of every user and every movie they rated, okay?

Now, we want to be able to work with movie titles, so we can interpret these results more intuitively, so we're going to use their human-readable names instead.

If you're using a truly massive dataset, you'd save that to the end because you want to be working with numbers, they're more compact, for as long as possible. For the purpose of example and teaching, though, we'll keep the titles around so you can see what's going on.

```
m_cols = ['movie_id', 'title']
movies = pd.read_csv('e:/sundog-consult/packt/datascience/ml-100k/u.item',
                sep='|', names=m_cols, usecols=range(2))
```

There's a separate data file with the MovieLens dataset called u.item, and it is pipe-delimited, and the first two columns that we import will be the movie_id and the title of that movie. So, now we have two DataFrames: r_cols has all the user ratings and m_cols has all the titles for every movie_id. We can then use the magical merge function in Pandas to mush it all together.

```
ratings = pd.merge(movies, ratings)
```

Let's add a `ratings.head()` command and then run those cells. What we end up with is something like the following table. That was pretty quick!

In [2]:	ratings.head()				
Out[2]:		movie_id	title	user_id	rating
	0	1	Toy Story (1995)	308	4
	1	1	Toy Story (1995)	287	5
	2	1	Toy Story (1995)	148	4
	3	1	Toy Story (1995)	280	4
	4	1	Toy Story (1995)	66	3

We end up with a new DataFrame that contains the `user_id` and rating for each movie that a user rated, and we have both the `movie_id` and the `title` that we can read and see what it really is. So, the way to read this is `user_id` number 308 rated the `Toy Story (1995)` movie 4 stars, `user_id` number 287 rated the `Toy Story (1995)` movie 5 stars, and so on and so forth. And, if we were to keep looking at more and more of this DataFrame, we'd see different ratings for different movies as we go through it.

Now the real magic of Pandas comes in. So, what we really want is to look at relationships between movies based on all the users that watched each pair of movies, so we need, at the end, a matrix of every movie, and every user, and all the ratings that every user gave to every movie. The `pivot_table` command in Pandas can do that for us. It can basically construct a new table from a given DataFrame, pretty much any way that you want it. For this, we can use the following code:

```
movieRatings = ratings.pivot_table(index=['user_id'],
                    columns=['title'],values='rating')
movieRatings.head()
```

So, what we're saying with this code is-take our ratings DataFrame and create a new DataFrame called `movieRatings` and we want the index of it to be the user IDs, so we'll have a row for every `user_id`, and we're going to have every column be the movie title. So, we're going to have a column for every title that we encounter in that DataFrame, and each cell will contain the `rating` value, if it exists. So, let's go ahead and run it.

And, we end up with a new DataFrame that looks like the following table:

title	'Til There Was You (1997)	1-900 (1994)	101 Dalmatians (1996)	12 Angry Men (1957)	187 (1997)	2 Days in the Valley (1996)	20,000 Leagues Under the Sea (1954)	2001: A Space Odyssey (1968)	3 Ninjas: High Noon At Mega Mountain (1998)	39 Steps, The (1935)	...	Yankee Zulu (1994)	Year of the Horse (1997)	You So Crazy (1994)	Young Frankenstein (1974)	Young Guns (1988)
user_id																
0	NaN	NaN	NaN	NaN	NaN	NaN	NaN	NaN	NaN	NaN	...	NaN	NaN	NaN	NaN	NaN
1	NaN	NaN	2	5	NaN	NaN	3	4	NaN	NaN	...	NaN	NaN	NaN	5	3
2	NaN	NaN	NaN	NaN	NaN	NaN	NaN	NaN	1	NaN	...	NaN	NaN	NaN	NaN	NaN
3	NaN	NaN	NaN	NaN	2	NaN	NaN	NaN	NaN	NaN	...	NaN	NaN	NaN	NaN	NaN
4	NaN	NaN	NaN	NaN	NaN	NaN	NaN	NaN	NaN	NaN	...	NaN	NaN	NaN	NaN	NaN

5 rows × 1664 columns

It's kind of amazing how that just put it all together for us. Now, you'll see some NaN values, which stands for **Not a Number**, and its just how Pandas indicates a missing value. So, the way to interpret this is, user_id number 1, for example, did not watch the movie 1-900 (1994), but user_id number 1 did watch 101 Dalmatians (1996) and rated it 2 stars. The user_id number 1 also watched 12 Angry Men (1957) and rated it 5 stars, but did not watch the movie 2 Days in the Valley (1996), for example, okay? So, what we end up with here is a sparse matrix basically, that contains every user, and every movie, and at every intersection where a user rated a movie there's a rating value.

So, you can see now, we can very easily extract vectors of every movie that our user watched, and we can also extract vectors of every user that rated a given movie, which is what we want. So, that's useful for both user-based and item-based collaborative filtering, right? If I wanted to find relationships between users, I could look at correlations between these user rows, but if I want to find correlations between movies, for item-based collaborative filtering, I can look at correlations between columns based on the user behavior. So, this is where the real *flipping things on its head for user versus item-based similarities* comes into play.

Now, we're going with item-based collaborative filtering, so we want to extract columns, to do this let's run the following code:

```
starWarsRatings = movieRatings['Star Wars (1977)']
starWarsRatings.head()
```

Now, with the help of that, let's go ahead and extract all the users who rated `Star Wars (1977)`:

```
Out[4]: user_id
        0      5
        1      5
        2      5
        3    NaN
        4      5
        Name: Star Wars (1977), dtype: float64
```

And, we can see most people have, in fact, watched and rated `Star Wars (1977)` and everyone liked it, at least in this little sample that we took from the head of the DataFrame. So, we end up with a resulting set of user IDs and their ratings for `Star Wars (1977)`. The user ID 3 did not rate `Star Wars (1977)` so we have a `NaN` value, indicating a missing value there, but that's okay. We want to make sure that we preserve those missing values so we can directly compare columns from different movies. So, how do we do that?

The corrwith function

Well, Pandas keeps making it easy for us, and has a `corrwith` function that you can see in the following code that we can use:

```
similarMovies = movieRatings.corrwith(starWarsRatings)
similarMovies = similarMovies.dropna()
df = pd.DataFrame(similarMovies)
df.head(10)
```

That code will go ahead and correlate a given column with every other column in the DataFrame, and compute the correlation scores and give that back to us. So, what we're doing here is using `corrwith` on the entire `movieRatings` DataFrame, that's that entire matrix of user movie ratings, correlating it with just the `starWarsRatings` column, and then dropping all of the missing results with `dropna`. So, that just leaves us with items that had a correlation, where there was more than one person that viewed it, and we create a new DataFrame based on those results and then display the top 10 results. So again, just to recap:

1. We're going to build the correlation score between Star Wars and every other movie.
2. Drop all the `NaN` values, so that we only have movie similarities that actually exist, where more than one person rated it.
3. And, we're going to construct a new DataFrame from the results and look at the top 10 results.

And here we are with the results shown in the following screenshot:

Out[5]:

title	0
'Til There Was You (1997)	0.872872
1-900 (1994)	-0.645497
101 Dalmatians (1996)	0.211132
12 Angry Men (1957)	0.184289
187 (1997)	0.027398
2 Days in the Valley (1996)	0.066654
20,000 Leagues Under the Sea (1954)	0.289768
2001: A Space Odyssey (1968)	0.230884
39 Steps, The (1935)	0.106453
8 1/2 (1963)	-0.142977

We ended up with this result of correlation scores between each individual movie for Star Wars and we can see, for example, a surprisingly high correlation score with the movie `'Til There Was You (1997)`, a negative correlation with the movie 1-900 (1994), and a very weak correlation with `101 Dalmatians (1996)`.

Now, all we should have to do is sort this by similarity score, and we should have the top movie similarities for Star Wars, right? Let's go ahead and do that.

```
similarMovies.sort_values(ascending=False)
```

Just call `sort_values` on the resulting DataFrame, again Pandas makes it really easy, and we can say `ascending=False`, to actually get it sorted in reverse order by correlation score. So, let's do that:

```
Out[6]: title
        Full Speed (1996)                                                              1.000000
        Star Wars (1977)                                                               1.000000
        Mondo (1996)                                                                   1.000000
        Man of the Year (1995)                                                         1.000000
        Line King: Al Hirschfeld, The (1996)                                           1.000000
        Outlaw, The (1943)                                                             1.000000
        Hurricane Streets (1998)                                                       1.000000
        Hollow Reed (1996)                                                             1.000000
        Scarlet Letter, The (1926)                                                     1.000000
        Safe Passage (1994)                                                            1.000000
        Good Man in Africa, A (1994)                                                   1.000000
        Golden Earrings (1947)                                                         1.000000
        Old Lady Who Walked in the Sea, The (Vieille qui marchait dans la mer, La) (1991)  1.000000
        No Escape (1994)                                                               1.000000
        Ed's Next Move (1996)                                                          1.000000
        Stripes (1981)                                                                 1.000000
        Cosi (1996)                                                                    1.000000
        Commandments (1997)                                                            1.000000
        Twisted (1996)                                                                 1.000000
        Beans of Egypt, Maine, The (1994)                                              1.000000
        Last Time I Saw Paris, The (1954)                                              1.000000
        Maya Lin: A Strong Clear Vision (1994)                                         1.000000
        Designated Mourner, The (1997)                                                 0.970725
        Albino Alligator (1996)                                                        0.968496
        Angel Baby (1995)                                                              0.962250
        Prisoner of the Mountains (Kavkazsky Plennik) (1996)                           0.927173
        Love in the Afternoon (1957)                                                   0.923381
        'Til There Was You (1997)                                                      0.872872
        A Chef in Love (1996)                                                          0.868599
```

Okay, so `Star Wars (1977)` came out pretty close to top, because it is similar to itself, but what's all this other stuff? What the heck? We can see in the preceding output, some movies such as: `Full Speed (1996)`, `Man of the Year (1995)`, `The Outlaw (1943)`. These are all, you know, fairly obscure movies, that most of them I've never even heard of, and yet they have perfect correlations with Star Wars. That's kinda weird! So, obviously we're doing something wrong here. What could it be?

Recommender Systems

Well, it turns out there's a perfectly reasonable explanation, and this is a good lesson in why you always need to examine your results when you're done with any sort of data science task-question the results, because often there's something you missed, there might be something you need to clean in your data, there might be something you did wrong. But you should also always look skeptically at your results, don't just take them on faith, okay? If you do so, you're going to get in trouble, because if I were to actually present these as recommendations to people who liked Star Wars, I would get fired. Don't get fired! Pay attention to your results! So, let's dive into what went wrong in our next section.

Improving the results of movie similarities

Let's figure out what went wrong with our movie similarities there. We went through all this exciting work to compute correlation scores between movies based on their user ratings vectors, and the results we got kind of sucked. So, just to remind you, we looked for movies that are similar to Star Wars using that technique, and we ended up with a bunch of weird recommendations at the top that had a perfect correlation.

And, most of them were very obscure movies. So, what do you think might be going on there? Well, one thing that might make sense is, let's say we have a lot of people watch Star Wars and some other obscure film. We'd end up with a good correlation between these two movies because they're tied together by Star Wars, but at the end of the day, do we really want to base our recommendations on the behavior of one or two people that watch some obscure movie?

Probably not! I mean yes, the two people in the world, or whatever it is, that watch the movie Full Speed, and both liked it in addition to Star Wars, maybe that is a good recommendation for them, but it's probably not a good recommendation for the rest of the world. We need to have some sort of confidence level in our similarities by enforcing a minimum boundary of how many people watched a given movie. We can't make a judgment that a given movie is good just based on the behavior of one or two people.

So, let's try to put that insight into action using the following code:

```
import numpy as np
movieStats = ratings.groupby('title').agg({'rating': [np.size, np.mean]})
movieStats.head()
```

What we're going to do is try to identify the movies that weren't actually rated by many people and we'll just throw them out and see what we get. So, to do that we're going to take our original ratings DataFrame and we're going to say `groupby('title')`, again Pandas has all sorts of magic in it. And, this will basically construct a new DataFrame that aggregates together all the rows for a given title into one row.

We can say that we want to aggregate specifically on the rating, and we want to show both the size, the number of ratings for each movie, and the mean average score, the mean rating for that movie. So, when we do that, we end up with something like the following:

Out[8]:

title	rating	
	size	mean
'Til There Was You (1997)	9	2.333333
1-900 (1994)	5	2.600000
101 Dalmatians (1996)	109	2.908257
12 Angry Men (1957)	125	4.344000
187 (1997)	41	3.024390

This is telling us, for example, for the movie `101 Dalmatians (1996)`, 109 people rated that movie and their average rating was 2.9 stars, so not that great of a score really! So, if we just eyeball this data, we can say okay well, movies that I consider obscure, like `187 (1997)`, had 41 ratings, but `101 Dalmatians (1996)`, I've heard of that, you know `12 Angry Men (1957)`, I've heard of that. It seems like there's sort of a natural cutoff value at around 100 ratings, where maybe that's the magic value where things start to make sense.

Let's go ahead and get rid of movies rated by fewer than 100 people, and yes, you know I'm kind of doing this intuitively at this point. As we'll talk about later, there are more principled ways of doing this, where you could actually experiment and do train/test experiments on different threshold values, to find the one that actually performs the best. But initially, let's just use our common sense and filter out movies that were rated by fewer than 100 people. Again, Pandas makes that really easy to do. Let's figure it out with the following example:

```
popularMovies = movieStats['rating']['size'] >= 100
movieStats[popularMovies].sort_values([('rating', 'mean')],
ascending=False)[:15]
```

We can just say `popularMovies`, a new DataFrame, is going to be constructed by looking at `movieStats` and we're going to only take rows where the rating size is greater than or equal to `100`, and I'm then going to sort that by `mean` rating, just for fun, to see the top rated, widely watched movies.

Out[9]:

title	rating size	rating mean
Close Shave, A (1995)	112	4.491071
Schindler's List (1993)	298	4.466443
Wrong Trousers, The (1993)	118	4.466102
Casablanca (1942)	243	4.456790
Shawshank Redemption, The (1994)	283	4.445230
Rear Window (1954)	209	4.387560
Usual Suspects, The (1995)	267	4.385768
Star Wars (1977)	584	4.359589
12 Angry Men (1957)	125	4.344000
Citizen Kane (1941)	198	4.292929
To Kill a Mockingbird (1962)	219	4.292237
One Flew Over the Cuckoo's Nest (1975)	264	4.291667
Silence of the Lambs, The (1991)	390	4.289744
North by Northwest (1959)	179	4.284916
Godfather, The (1972)	413	4.283293

What we have here is a list of movies that were rated by more than 100 people, sorted by their average rating score, and this in itself is a recommender system. These are highly-rated popular movies. `A Close Shave (1995)`, apparently, was a really good movie and a lot of people watched it and they really liked it.

So again, this is a very old dataset, from the late 90s, so even though you might not be familiar with the film `A Close Shave (1995)`, it might be worth going back and rediscovering it; add it to your Netflix! `Schindler's List (1993)` not a big surprise there, that comes up on the top of most top movies lists. `The Wrong Trousers (1993)`, another example of an obscure film that apparently was really good and was also pretty popular. So, some interesting discoveries there already, just by doing that.

Things look a little bit better now, so let's go ahead and basically make our new DataFrame of Star Wars recommendations, movies similar to Star Wars, where we only base it on movies that appear in this new DataFrame. So, we're going to use the `join` operation, to go ahead and join our original `similarMovies` DataFrame to this new DataFrame of only movies that have greater than 100 ratings, okay?

```
df = movieStats[popularMovies].join(pd.DataFrame(similarMovies,
columns=['similarity']))
df.head()
```

In this code, we create a new DataFrame based on `similarMovies` where we extract the `similarity` column, join that with our `movieStats` DataFrame, which is our `popularMovies` DataFrame, and we look at the combined results. And, there we go with that output!

Out[11]:

title	(rating, size)	(rating, mean)	similarity
101 Dalmatians (1996)	109	2.908257	0.211132
12 Angry Men (1957)	125	4.344000	0.184289
2001: A Space Odyssey (1968)	259	3.969112	0.230884
Absolute Power (1997)	127	3.370079	0.085440
Abyss, The (1989)	151	3.589404	0.203709

Now we have, restricted only to movies that are rated by more than 100 people, the similarity score to Star Wars. So, now all we need to do is sort that using the following code:

```
df.sort_values(['similarity'], ascending=False)[:15]
```

Recommender Systems

Here, we're reverse sorting it and we're just going to take a look at the first 15 results. If you run that now, you should see the following:

Out[12]:

title	(rating, size)	(rating, mean)	similarity
Star Wars (1977)	584	4.359589	1.000000
Empire Strikes Back, The (1980)	368	4.206522	0.748353
Return of the Jedi (1983)	507	4.007890	0.672556
Raiders of the Lost Ark (1981)	420	4.252381	0.536117
Austin Powers: International Man of Mystery (1997)	130	3.246154	0.377433
Sting, The (1973)	241	4.058091	0.367538
Indiana Jones and the Last Crusade (1989)	331	3.930514	0.350107
Pinocchio (1940)	101	3.673267	0.347868
Frighteners, The (1996)	115	3.234783	0.332729
L.A. Confidential (1997)	297	4.161616	0.319065
Wag the Dog (1997)	137	3.510949	0.318645
Dumbo (1941)	123	3.495935	0.317656
Bridge on the River Kwai, The (1957)	165	4.175758	0.316580
Philadelphia Story, The (1940)	104	4.115385	0.314272
Miracle on 34th Street (1994)	101	3.722772	0.310921

This is starting to look a little bit better! So, `Star Wars (1977)` comes out on top because it's similar to itself, `The Empire Strikes Back (1980)` is number 2, `Return of the Jedi (1983)` is number 3, `Raiders of the Lost Ark (1981)`, number 4. You know, it's still not perfect, but these make a lot more sense, right? So, you would expect the three Star Wars films from the original trilogy to be similar to each other, this data goes back to before the next three films, and `Raiders of the Lost Ark (1981)` is also a very similar movie to Star Wars in style, and it comes out as number 4. So, I'm starting to feel a little bit better about these results. There's still room for improvement, but hey! We got some results that make sense, whoo-hoo!

Now, ideally, we'd also filter out Star Wars, you don't want to be looking at similarities to the movie itself that you started from, but we'll worry about that later! So, if you want to play with this a little bit more, like I said 100 was sort of an arbitrary cutoff for the minimum number of ratings. If you do want to experiment with different cutoff values, I encourage you to go back and do so. See what that does to the results. You know, you can see in the preceding table that the results that we really like actually had much more than 100 ratings in common. So, we end up with `Austin Powers: International Man of Mystery (1997)` coming in there pretty high with only `130` ratings so maybe 100 isn't high enough! `Pinocchio (1940)` snuck in at `101`, not very similar to Star Wars, so, you might want to consider an even higher threshold there and see what it does.

 Please keep in mind too, this is a very small, limited dataset that we used for experimentation purposes, and it's based on very old data, so you're only going to see older movies. So, interpreting these results intuitively might be a little bit challenging as a result, but not bad results.

Now let's move on and actually do full-blown item-based collaborative filtering where we recommend movies to people using a more complete system, we'll do that next.

Making movie recommendations to people

Okay, let's actually build a full-blown recommender system that can look at all the behavior information of everybody in the system, and what movies they rated, and use that to actually produce the best recommendation movies for any given user in our dataset. Kind of amazing and you'll be surprised how simple it is. Let's go!

Let's begin using the `ItemBasedCF.ipynb` file and let's start off by importing the MovieLens dataset that we have. Again, we're using a subset of it that just contains 100,000 ratings for now. But, there are larger datasets you can get from GroupLens.org-up to millions of ratings; if you're so inclined. Keep in mind though, when you start to deal with that really big data, you're going to be pushing the limits of what you can do in a single machine and what Pandas can handle. Without further ado, here's the first block of code:

```
import pandas as pd

r_cols = ['user_id', 'movie_id', 'rating']
ratings = pd.read_csv('e:/sundog-consult/packt/datascience/ml-100k/u.data',
                      sep='\t', names=r_cols, usecols=range(3))

m_cols = ['movie_id', 'title']
movies = pd.read_csv('e:/sundog-consult/packt/datascience/ml-100k/u.item',
                     sep='|', names=m_cols, usecols=range(2))
```

Recommender Systems

```
ratings = pd.merge(movies, ratings)

ratings.head()
```

Just like earlier, we're going to import the `u.data` file that contains all the individual ratings for every user and what movie they rated, and then we're going to tie that together with the movie titles, so we don't have to just work with numerical movie IDs. Go ahead and hit the run cell button, and we end up with the following DataFrame.

Out[1]:

	movie_id	title	user_id	rating
0	1	Toy Story (1995)	308	4
1	1	Toy Story (1995)	287	5
2	1	Toy Story (1995)	148	4
3	1	Toy Story (1995)	280	4
4	1	Toy Story (1995)	66	3

The way to read this is, for example, `user_id` number 308 rated `Toy Story (1995)` a 4 star, and `user_id` number 66 rated `Toy Story (1995)` a 3 star. And, this will contain every rating, for every user, for every movie.

And again, just like earlier, we use the wonderful `pivot_table` command in Pandas to construct a new DataFrame based on the information:

```
userRatings = ratings.pivot_table(index=['user_id'],
                    columns=['title'],values='rating')
userRatings.head()
```

Here, each row is the `user_id`, the columns are made up of all the unique movie titles in my dataset, and each cell contains a rating:

title	'Til There Was You (1997)	1-900 (1994)	101 Dalmatians (1996)	12 Angry Men (1957)	187 (1997)	2 Days in the Valley (1996)	20,000 Leagues Under the Sea (1954)	2001: A Space Odyssey (1968)	3 Ninjas: High Noon At Mega Mountain (1998)	39 Steps, The (1935)	...	Yankee Zulu (1994)	Year of the Horse (1997)	You So Crazy (1994)	Young Frankenstein (1974)	Young Guns (1988)
user_id																
0	NaN	NaN	NaN	NaN	NaN	NaN	NaN	NaN	NaN	NaN	...	NaN	NaN	NaN	NaN	NaN
1	NaN	NaN	2	5	NaN	3	4	NaN	NaN	NaN	...	NaN	NaN	NaN	5	3
2	NaN	NaN	NaN	NaN	NaN	NaN	NaN	1	NaN	NaN	...	NaN	NaN	NaN	NaN	NaN
3	NaN	NaN	NaN	NaN	2	NaN	NaN	NaN	NaN	NaN	...	NaN	NaN	NaN	NaN	NaN
4	NaN	NaN	NaN	NaN	NaN	NaN	NaN	NaN	NaN	NaN	...	NaN	NaN	NaN	NaN	NaN

5 rows × 1664 columns

What we end up with is this incredibly useful matrix shown in the preceding output, that contains users for every row and movies for every column. And we have basically every user rating for every movie in this matrix. So, `user_id` number 1, for example, gave `101 Dalmatians (1996)` a 2-star rating. And, again all these `NaN` values represent missing data. So, that just indicates, for example, `user_id` number 1 did not rate the movie `1-900 (1994)`.

Again, it's a very useful matrix to have. If we were doing user-based collaborative filtering, we could compute correlations between each individual user rating vector to find similar users. Since we're doing item-based collaborative filtering, we're more interested in relationships between the columns. So, for example, doing a correlation score between any two columns, which will give us a correlation score for a given movie pair. So, how do we do that? It turns out that Pandas makes that incredibly easy to do as well.

It has a built-in `corr` function that will actually compute the correlation score for every column pair found in the entire matrix-it's almost like they were thinking of us.

```
corrMatrix = userRatings.corr()
corrMatrix.head()
```

Let's go ahead and run the preceding code. It's a fairly computationally expensive thing to do, so it will take a moment to actually come back with a result. But, there we have it!

title	'Til There Was You (1997)	1-900 (1994)	101 Dalmatians (1996)	12 Angry Men (1957)	187 (1997)	2 Days in the Valley (1996)	20,000 Leagues Under the Sea (1954)	2001: A Space Odyssey (1968)	3 Ninjas: High Noon At Mega Mountain (1998)	39 Steps, The (1935)	...	Yankee Zulu (1994)	Year of the Horse (1997)	You So Crazy (1994)
title														
'Til There Was You (1997)	1.0	NaN	-1.000000	-0.500000	-0.500000	0.522233	NaN	-0.426401	NaN	NaN	...	NaN	NaN	NaN
1-900 (1994)	NaN	1	NaN	NaN	NaN	NaN	NaN	-0.981981	NaN	NaN	...	NaN	NaN	NaN
101 Dalmatians (1996)	-1.0	NaN	1.000000	-0.049890	0.269191	0.048973	0.266928	-0.043407	NaN	0.111111	...	NaN	-1.000000	NaN
12 Angry Men (1957)	-0.5	NaN	-0.049890	1.000000	0.666667	0.256625	0.274772	0.178848	NaN	0.457176	...	NaN	NaN	NaN
187 (1997)	-0.5	NaN	0.269191	0.666667	1.000000	0.596644	NaN	-0.554700	NaN	1.000000	...	NaN	0.866025	NaN

5 rows × 1664 columns

So, what do we have in the preceding output? We have here a new DataFrame where every movie is on the row, and in the column. So, we can look at the intersection of any two given movies and find their correlation score to each other based on this `userRatings` data that we had up here originally. How cool is that? For example, the movie `101 Dalmatians (1996)` is perfectly correlated with itself of course, because it has identical user rating vectors. But, if you look at `101 Dalmatians (1996)` movie's relationship to the movie `12 Angry Men (1957)`, it's a much lower correlation score because those movies are rather dissimilar, makes sense, right?

I have this wonderful matrix now that will give me the similarity score of any two movies to each other. It's kind of amazing, and very useful for what we're going to be doing. Now just like earlier, we have to deal with spurious results. So, I don't want to be looking at relationships that are based on a small amount of behavior information.

It turns out that the Pandas `corr` function actually has a few parameters you can give it. One is the actual correlation score method that you want to use, so I'm going to say use `pearson` correlation.

```
corrMatrix = userRatings.corr(method='pearson', min_periods=100)
corrMatrix.head()
```

You'll notice that it also has a `min_periods` parameter you can give it, and that basically says I only want you to consider correlation scores that are backed up by at least, in this example, 100 people that rated both movies. Running that will get rid of the spurious relationships that are based on just a handful of people. The following is the matrix that we get after running the code:

Out[4]: title	'Til There Was You (1997)	1-900 (1994)	101 Dalmatians (1996)	12 Angry Men (1957)	187 (1997)	2 Days in the Valley (1996)	20,000 Leagues Under the Sea (1954)	2001: A Space Odyssey (1968)	3 Ninjas: High Noon At Mega Mountain (1998)	39 Steps, The (1935)	...	Yankee Zulu (1994)	Year of the Horse (1997)	You So Crazy (1994)	Young Frankenstein (1974)
title															
'Til There Was You (1997)	NaN	NaN	NaN	NaN	NaN	NaN	NaN	NaN	NaN	NaN	...	NaN	NaN	NaN	NaN
1-900 (1994)	NaN	NaN	NaN	NaN	NaN	NaN	NaN	NaN	NaN	NaN	...	NaN	NaN	NaN	NaN
101 Dalmatians (1996)	NaN	NaN	1	NaN	NaN	NaN	NaN	NaN	NaN	NaN	...	NaN	NaN	NaN	NaN
12 Angry Men (1957)	NaN	NaN	NaN	1	NaN	NaN	NaN	NaN	NaN	NaN	...	NaN	NaN	NaN	NaN
187 (1997)	NaN	NaN	NaN	NaN	NaN	NaN	NaN	NaN	NaN	NaN	...	NaN	NaN	NaN	NaN

5 rows × 1664 columns

It's a little bit different to what we did in the item similarities exercise where we just threw out any movie that was rated by less than 100 people. What we're doing here, is throwing out movie similarities where less than 100 people rated both of those movies, okay? So, you can see in the preceding matrix that we have a lot more `NaN` values.

In fact, even movies that are similar to themselves get thrown out, so for example, the movie `1-900 (1994)` was, presumably, watched by fewer than 100 people so it just gets tossed entirely. The movie, `101 Dalmatians (1996)` however, survives with a correlation score of 1, and there are actually no movies in this little sample of the dataset that are different from each other that had 100 people in common that watched both. But, there are enough movies that survive to get meaningful results.

[237]

Understanding movie recommendations with an example

So, what we do with this data? Well, what we want to do is recommend movies for people. The way we do that is we look at all the ratings for a given person, find movies similar to the stuff that they rated, and those are candidates for recommendations to that person.

Let's start by creating a fake person to create recommendations for. I've actually already added a fake user by hand, ID number 0, to the MovieLens dataset that we're processing. You can see that user with the following code:

```
myRatings = userRatings.loc[0].dropna()
myRatings
```

This gives the following output:

```
Out[5]:  title
         Empire Strikes Back, The (1980)    5
         Gone with the Wind (1939)          1
         Star Wars (1977)                   5
         Name: 0, dtype: float64
```

That kind of represents someone like me, who loved Star Wars and The Empire Strikes Back, but hated the movie Gone with the Wind. So, this represents someone who really loves Star Wars, but does not like old style, romantic dramas, okay? So, I gave a rating of 5 star to `The Empire Strikes Back (1980)` and `Star Wars (1977)`, and a rating of 1 star to `Gone with the Wind (1939)`. So, I'm going to try to find recommendations for this fictitious user.

So, how do I do that? Well, let's start by creating a series called `simCandidates` and I'm going to go through every movie that I rated.

```
simCandidates = pd.Series()
for i in range(0, len(myRatings.index)):
    print "Adding sims for " + myRatings.index[i] + "..."
    # Retrieve similar movies to this one that I rated
    sims = corrMatrix[myRatings.index[i]].dropna()
    # Now scale its similarity by how well I rated this movie
    sims = sims.map(lambda x: x * myRatings[i])
    # Add the score to the list of similarity candidates
    simCandidates = simCandidates.append(sims)
```

```
#Glance at our results so far:
print "sorting..."
simCandidates.sort_values(inplace = True, ascending = False)
print simCandidates.head(10)
```

For i in range 0 through the number of ratings that I have in myRatings, I am going to add up similar movies to the ones that I rated. So, I'm going to take that corrMatrix DataFrame, that magical one that has all of the movie similarities, and I am going to create a correlation matrix with myRatings, drop any missing values, and then I am going to scale that resulting correlation score by how well I rated that movie.

So, the idea here is I'm going to go through all the similarities for The Empire Strikes Back, for example, and I will scale it all by 5, because I really liked The Empire Strikes Back. But, when I go through and get the similarities for Gone with the Wind, I'm only going to scale those by 1, because I did not like Gone with the Wind. So, this will give more strength to movies that are similar to movies that I liked, and less strength to movies that are similar to movies that I did not like, okay?

So, I just go through and build up this list of similarity candidates, recommendation candidates if you will, sort the results and print them out. Let's see what we get:

```
Adding sims for Empire Strikes Back, The (1980)...
Adding sims for Gone with the Wind (1939)...
Adding sims for Star Wars (1977)...
sorting...
title
Empire Strikes Back, The (1980)                         5.000000
Star Wars (1977)                                        5.000000
Empire Strikes Back, The (1980)                         3.741763
Star Wars (1977)                                        3.741763
Return of the Jedi (1983)                               3.606146
Return of the Jedi (1983)                               3.362779
Raiders of the Lost Ark (1981)                          2.693297
Raiders of the Lost Ark (1981)                          2.680586
Austin Powers: International Man of Mystery (1997)      1.887164
Sting, The (1973)                                       1.837692
dtype: float64
```

Hey, those don't look too bad, right? So, obviously The Empire Strikes Back (1980) and Star Wars (1977) come out on top, because I like those movies explicitly, I already watched them and rated them. But, bubbling up to the top of the list is Return of the Jedi (1983), which we would expect and Raiders of the Lost Ark (1981).

Let's start to refine these results a little bit more. We're seeing that we're getting duplicate values back. If we have a movie that was similar to more than one movie that I rated, it will come back more than once in the results, so we want to combine those together. If I do in fact have the same movie, maybe that should get added up together into a combined, stronger recommendation score. Return of the Jedi, for example, was similar to both Star Wars and The Empire Strikes Back. How would we do that?

Using the groupby command to combine rows

We'll go ahead and explore that. We're going to use the `groupby` command again to group together all of the rows that are for the same movie. Next, we will sum up their correlation score and look at the results:

```
simCandidates = simCandidates.groupby(simCandidates.index).sum()
simCandidates.sort_values(inplace = True, ascending = False)
simCandidates.head(10)
```

Following is the result:

```
Out[8]:  title
         Empire Strikes Back, The (1980)              8.877450
         Star Wars (1977)                             8.870971
         Return of the Jedi (1983)                    7.178172
         Raiders of the Lost Ark (1981)               5.519700
         Indiana Jones and the Last Crusade (1989)    3.488028
         Bridge on the River Kwai, The (1957)         3.366616
         Back to the Future (1985)                    3.357941
         Sting, The (1973)                            3.329843
         Cinderella (1950)                            3.245412
         Field of Dreams (1989)                       3.222311
         dtype: float64
```

Hey, this is looking really good!

So `Return of the Jedi (1983)` comes out way on top, as it should, with a score of 7, `Raiders of the Lost Ark (1981)` a close second at 5, and then we start to get to `Indiana Jones and the Last Crusade (1989)`, and some more movies, `The Bridge on the River Kwai (1957)`, `Back to the Future (1985)`, `The Sting (1973)`. These are all movies that I would actually enjoy watching! You know, I actually do like old-school Disney movies too, so `Cinderella (1950)` isn't as crazy as it might seem.

The last thing we need to do is filter out the movies that I've already rated, because it doesn't make sense to recommend movies you've already seen.

Removing entries with the drop command

So, I can quickly drop any rows that happen to be in my original ratings series using the following code:

```
filteredSims = simCandidates.drop(myRatings.index)
filteredSims.head(10)
```

Running that will let me see the final top 10 results:

```
Out[9]: title
        Return of the Jedi (1983)                      7.178172
        Raiders of the Lost Ark (1981)                 5.519700
        Indiana Jones and the Last Crusade (1989)      3.488028
        Bridge on the River Kwai, The (1957)           3.366616
        Back to the Future (1985)                      3.357941
        Sting, The (1973)                              3.329843
        Cinderella (1950)                              3.245412
        Field of Dreams (1989)                         3.222311
        Wizard of Oz, The (1939)                       3.200268
        Dumbo (1941)                                   2.981645
        dtype: float64
```

And there we have it! `Return of the Jedi (1983)`, `Raiders of the Lost Ark (1981)`, `Indiana Jones and the Last Crusade (1989)`, all the top results for my fictitious user, and they all make sense. I'm seeing a few family-friendly films, you know, `Cinderella (1950)`, `The Wizard of Oz (1939)`, `Dumbo (1941)`, creeping in, probably based on the presence of Gone with the Wind in there, even though it was weighted downward it's still in there, and still being counted. And, there we have our results, so. There you have it! Pretty cool!

We have actually generated recommendations for a given user and we could do that for any user in our entire DataFrame. So, go ahead and play with that if you want to. I also want to talk about how you can actually get your hands dirty a little bit more, and play with these results; try to improve upon them.

There's a bit of an art to this, you know, you need to keep iterating and trying different ideas and different techniques until you get better and better results, and you can do this pretty much forever. I mean, I made a whole career out of it. So, I don't expect you to spend the next, you know, 10 years trying to refine this like I did, but there are some simple things you can do, so let's talk about that.

Improving the recommendation results

As an exercise, I want to challenge you to go and make those recommendations even better. So, let's talk about some ideas I have, and maybe you'll have some of your own too that you can actually try out and experiment with; get your hands dirty, and try to make better movie recommendations.

Okay, there's a lot of room for improvement still on these recommendation results. There's a lot of decisions we made about how to weigh different recommendation results based on your rating of that item that it came from, or what threshold you want to pick for the minimum number of people that rated two given movies. So, there's a lot of things you can tweak, a lot of different algorithms you can try, and you can have a lot of fun with trying to make better movie recommendations out of the system. So, if you're feeling up to it, I'm challenging you to go and do just that!

Here are some ideas on how you might actually try to improve upon the results in this chapter. First, you can just go ahead and play with the `ItembasedCF.ipynb` file and tinker with it. So, for example, we saw that the correlation method actually had some parameters for the correlation computation, we used Pearson in our example, but there are other ones you can look up and try out, see what it does to your results. We used a minimum period value of 100, maybe that's too high, maybe it's too low; we just kind of picked it arbitrarily. What happens if you play with that value? If you were to lower that for example, I would expect you to see some new movies maybe you've never heard of, but might still be a good recommendation for that person. Or, if you were to raise it higher, you would see, you know nothing but blockbusters.

Sometimes you have to think about what the result is that you want out of a recommender system. Is there a good balance to be had between showing people movies that they've heard of and movies that they haven't heard of? How important is discovery of new movies to these people versus having confidence in the recommender system by seeing a lot of movies that they have heard of? So again, there's sort of an art to that.

We can also improve upon the fact that we saw a lot of movies in the results that were similar to Gone with the Wind, even though I didn't like Gone with the Wind. You know we weighted those results lower than similarities to movies that I enjoyed, but maybe those movies should actually be penalized. If I hated Gone with the Wind that much, maybe similarities to Gone with the Wind, like The Wizard of Oz, should actually be penalized and, you know lowered in their score instead of raised at all.

That's another simple modification you can make and play around with. There are probably some outliers in our user rating dataset, what if I were to throw away people that rated some ridiculous number of movies? Maybe they're skewing everything. You could actually try to identify those users and throw them out, as another idea. And, if you really want a big project, if you really want to sink your teeth into this stuff, you could actually evaluate the results of this recommender engine by using the techniques of train/test. So, what if instead of having an arbitrary recommendation score that sums up the correlation scores of each individual movie, actually scale that down to a predicted rating for each given movie.

If the output of my recommender system were a movie and my predicted rating for that movie, in a train/test system I could actually try to figure out how well do I predict movies that the user has in fact watched and rated before? Okay? So, I could set aside some of the ratings data and see how well my recommender system is able to predict the user's ratings for those movies. And, that would be a quantitative and principled way to measure the error of this recommender engine. But again, there's a little bit more of an art than a science to this. Even though the Netflix prize actually used that error metric, called root-mean-square error is what they used in particular, is that really a measure of a good recommender system?

Basically, you're measuring the ability of your recommender system to predict the ratings of movies that a person already watched. But isn't the purpose of a recommender engine to recommend movies that a person hasn't watched, that they might enjoy? Those are two different things. So unfortunately, it's not very easy to measure the thing you really want to be measuring. So sometimes, you do kind of have to go with your gut instinct. And, the right way to measure the results of a recommender engine is to measure the results that you're trying to promote through it.

Maybe I'm trying to get people to watch more movies, or rate new movies more highly, or buy more stuff. Running actual controlled experiments on a real website would be the right way to optimize for that, as opposed to using train/test. So, you know, I went into a little bit more detail there than I probably should have, but the lesson is, you can't always think about these things in black and white. Sometimes, you can't really measure things directly and quantitatively, and you have to use a little bit of common sense, and this is an example of that.

Anyway, those are some ideas on how to go back and improve upon the results of this recommender engine that we wrote. So, please feel free to tinker around with it, see if you can improve upon it however you wish to, and have some fun with it. This is actually a very interesting part of the book, so I hope you enjoy it!

Summary

So, go give it a try! See if you can improve on our initial results there. There's some simple ideas there to try to make those recommendations better, and some much more complicated ones too. Now, there's no right or wrong answer; I'm not going to ask you to turn in your work, and I'm not going to review your work. You know, you decide to play around with it and get some familiarity with it, and experiment, and see what results you get. That's the whole point - just to get you more familiar with using Python for this sort of thing, and get more familiar with the concepts behind item-based collaborative filtering.

We've looked at different recommender systems in this chapter-we ruled out a user-based collaborative filtering system and dove straight in to an item-based system. We then used various functions from pandas to generate and refine our results, and I hope you've seen the power of pandas here.

In the next chapter, we'll take a look at more advanced data mining and machine learning techniques including K-nearest neighbors. I look forward to explaining those to you and seeing how they can be useful.

7
More Data Mining and Machine Learning Techniques

In this chapter, we talk about a few more data mining and machine learning techniques. We will talk about a really simple technique called **k-nearest neighbors** (**KNN**). We'll then use KNN to predict a rating for a movie. After that, we'll go on to talk about dimensionality reduction and principal component analysis. We'll also look at an example of PCA where we will reduce 4D data to two dimensions while still preserving its variance.

We'll then walk through the concept of data warehousing and see the advantages of the newer ELT process over the ETL process. We'll learn the fun concept of reinforcement learning and see the technique used behind the intelligent Pac-Man agent of the Pac-Man game. Lastly, we'll see some fancy terminology used for reinforcement learning.

We'll cover the following topics:

- The concept of k-nearest neighbors
- Implementation of KNN to predict the rating of a movie
- Dimensionality reduction and principal component analysis
- Example of PCA with the Iris dataset
- Data warehousing and ETL versus ELT
- What is reinforcement learning
- The working behind the intelligent Pac-Man game
- Some fancy words used for reinforcement learning

K-nearest neighbors - concepts

Let's talk about a few data mining and machine learning techniques that employers expect you to know about. We'll start with a really simple one called KNN for short. You're going to be surprised at just how simple a good supervised machine learning technique can be. Let's take a look!

KNN sounds fancy but it's actually one of the simplest techniques out there! Let's say you have a scatter plot and you can compute the distance between any two points on that scatter plot. Let's assume that you have a bunch of data that you've already classified, that you can train the system from. If I have a new data point, all I do is look at the KNN based on that distance metric and let them all vote on the classification of that new point.

Let's imagine that the following scatter plot is plotting movies. The squares represent science fiction movies, and the triangles represent drama movies. We'll say that this is plotting ratings versus popularity, or anything else you can dream up:

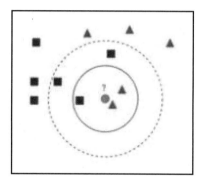

Here, we have some sort of distance that we can compute based on rating and popularity between any two points on the scatter plot. Let's say a new point comes in, a new movie that we don't know the genre for. What we could do is set K to 3 and take the 3 nearest neighbors to this point on the scatter plot; they can all then vote on the classification of the new point/movie.

You can see if I take the three nearest neighbors ($K=3$), I have 2 drama movies and 1 science fiction movie. I would then let them all vote, and we would choose the classification of drama for this new point based on those 3 nearest neighbors. Now, if I were to expand this circle to include 5 nearest neighbors, that is $K=5$, I get a different answer. So, in that case I pick up 3 science fiction and 2 drama movies. If I let them all vote I would end up with a classification of science fiction for the new movie instead.

Our choice of K can be very important. You want to make sure it's small enough that you don't go too far and start picking up irrelevant neighbors, but it has to be big enough to enclose enough data points to get a meaningful sample. So, often you'll have to use train/test or a similar technique to actually determine what the right value of *K* is for a given dataset. But, at the end of the day, you have to just start with your intuition and work from there.

That's all there is to it, it's just that simple. So, it is a very simple technique. All you're doing is literally taking the k nearest neighbors on a scatter plot, and letting them all vote on a classification. It does qualify as supervised learning because it is using the training data of a set of known points, that is, known classifications, to inform the classification of a new point.

But let's do something a little bit more complicated with it and actually play around with movies, just based on their metadata. Let's see if we can actually figure out the nearest neighbors of a movie based on just the intrinsic values of those movies, for example, the ratings for it, the genre information for it:

In theory, we could recreate something similar to *Customers Who Watched This Item Also Watched* (the above image is a screenshot from Amazon) just using k-nearest Neighbors. And, I can take it one step further: once I identify the movies that are similar to a given movie based on the k-nearest Neighbors algorithm, I can let them all vote on a predicted rating for that movie.

That's what we're going to do in our next example. So you now have the concepts of KNN, k-nearest neighbors. Let's go ahead and apply that to an example of actually finding movies that are similar to each other and using those nearest neighbor movies to predict the rating for another movie we haven't seen before.

Using KNN to predict a rating for a movie

Alright, we're going to actually take the simple idea of KNN and apply that to a more complicated problem, and that's predicting the rating of a movie given just its genre and rating information. So, let's dive in and actually try to predict movie ratings just based on the KNN algorithm and see where we get. So, if you want to follow along, go ahead and open up the `KNN.ipynb` and you can play along with me.

What we're going to do is define a distance metric between movies just based on their metadata. By metadata I just mean information that is intrinsic to the movie, that is, the information associated with the movie. Specifically, we're going to look at the genre classifications of the movie.

Every movie in our `MovieLens` dataset has additional information on what genre it belongs to. A movie can belong to more than one genre, a genre being something like science fiction, or drama, or comedy, or animation. We will also look at the overall popularity of the movie, given by the number of people who rated it, and we also know the average rating of each movie. I can combine all this information together to basically create a metric of distance between two movies just based on rating information and genre information. Let's see what we get.

We'll use pandas again to make life simple, and if you are following along, again make sure to change the path to the `MovieLens` dataset to wherever you installed it, which will almost certainly not be what is in this Python notebook.

Please go ahead and change that if you want to follow along. As before, we're just going to import the actual ratings data file itself, which is `u.data` using the `read_csv()` function in pandas. We're going to tell that it actually has a tab-delimiter and not a comma. We're going to import the first 3 columns, which represent the `user_id`, `movie_id`, and rating, for every individual movie rating in our dataset:

```
import pandas as pd

r_cols = ['user_id', 'movie_id', 'rating']
ratings = pd.read_csv('C:\DataScience\ml-100k\u.data', sep='\t',
names=r_cols, usecols=range(3))
ratings.head()ratings.head()
```

If we go ahead and run that and look at the top of it, we can see that it's working, here's how the output should look like:

	user_id	movie_id	rating
0	0	50	5
1	0	172	5
2	0	133	1
3	196	242	3
4	186	302	3

Out[1]:

We end up with a `DataFrame` that has `user_id`, `movie_id`, and `rating`. For example, `user_id` 0 rated `movie_id` 50, which I believe is Star Wars, 5 stars, and so on and so forth.

The next thing we have to figure out is aggregate information about the ratings for each movie. We use the `groupby()` function in pandas to actually group everything by `movie_id`. We're going to combine together all the ratings for each individual movie, and we're going to output the number of ratings and the average rating score, that is the mean, for each movie:

```
movieProperties = ratings.groupby('movie_id').agg({'rating':
    [np.size, np.mean]})
movieProperties.head()
```

Let's go ahead and do that - comes back pretty quickly, here's how the output looks like:

Out[3]:

	rating	
	size	mean
movie_id		
1	452	3.878319
2	131	3.206107
3	90	3.033333
4	209	3.550239
5	86	3.302326

More Data Mining and Machine Learning Techniques

This gives us another `DataFrame` that tells us, for example, `movie_id 1` had **452** ratings (which is a measure of its popularity, that is, how many people actually watched it and rated it), and a mean review score of 3.8. So, **452** people watched `movie_id 1`, and they gave it an average review of 3.87, which is pretty good.

Now, the raw number of ratings isn't that useful to us. I mean I don't know if **452** means it's popular or not. So, to normalize that, what we're going to do is basically measure that against the maximum and minimum number of ratings for each movie. We can do that using the `lambda` function. So, we can apply a function to an entire `DataFrame` this way.

What we're going to do is use the `np.min()` and `np.max()` functions to find the maximum number of ratings and the minimum number of ratings found in the entire dataset. So, we'll take the most popular movie and the least popular movie and find the range there, and normalize everything against that range:

```
movieNumRatings = pd.DataFrame(movieProperties['rating']['size'])
movieNormalizedNumRatings = movieNumRatings.apply(lambda x: (x - np.min(x))
/ (np.max(x) - np.min(x)))
movieNormalizedNumRatings.head()
```

What this gives us, when we run it, is the following:

Out[4]:

movie_id	size
1	0.773585
2	0.222985
3	0.152659
4	0.356775
5	0.145798

This is basically a measure of popularity for each movie, on a scale of 0 to 1. So, a score of 0 here would mean that nobody watched it, it's the least popular movie, and a score of 1 would mean that everybody watched, it's the most popular movie, or more specifically, the movie that the most people watched. So, we have a measure of movie popularity now that we can use for our distance metric.

Next, let's extract some general information. So, it turns out that there is a u.item file that not only contains the movie names, but also all the genres that each movie belongs to:

```
movieDict = {}
with open(r'c:/DataScience/ml-100k/u.item') as f:
    temp = ''
    for line in f:
        fields = line.rstrip('\n').split('|')
        movieID = int(fields[0])
        name = fields[1]
        genres = fields[5:25]
        genres = map(int, genres)
        movieDict[movieID] = (name, genres, movieNormalizedNumRatings.loc[movieID].get('size'),movieProperties.loc[movieID].rating.get('mean'))
```

The code above will actually go through each line of u.item. We're doing this the hard way; we're not using any pandas functions; we're just going to use straight-up Python this time. Again, make sure you change that path to wherever you installed this information.

Next, we open our u.item file, and then iterate through every line in the file one at a time. We strip out the new line at the end and split it based on the pipe-delimiters in this file. Then, we extract the movieID, the movie name and all of the individual genre fields. So basically, there's a bunch of 0s and 1s in 19 different fields in this source data, where each one of those fields represents a given genre. We then construct a Python dictionary in the end that maps movie IDs to their names, genres, and then we also fold back in our rating information. So, we will have name, genre, popularity on a scale of 0 to 1, and the average rating. So, that's what this little snippet of code does. Let's run that! And, just to see what we end up with, we can extract the value for movie_id 1:

```
movieDict[1]
```

Following is the output of the preceding code:

```
('Toy Story (1995)',
 [0, 0, 0, 1, 1, 1, 0, 0, 0, 0, 0, 0, 0, 0, 0, 0, 0, 0, 0],
 0.77358490566037741,
 3.8783185840707963)
```

Entry 1 in our dictionary for movie_id 1 happens to be Toy Story, an old Pixar film from 1995 you've probably heard of. Next is a list of all the genres, where a 0 indicates it is not part of that genre, and 1 indicates it is part of that genre. There is a data file in the MovieLens dataset that will tell you what each of these genre fields actually corresponds to.

For our purposes, that's not actually important, right? We're just trying to measure distance between movies based on their genres. So, all that matters mathematically is how similar this vector of genres is to another movie, okay? The actual genres themselves, not important! We just want to see how same or different two movies are in their genre classifications. So we have that genre list, we have the popularity score that we computed, and we have there the mean or average rating for Toy Story. Okay, let's go ahead and figure out how to combine all this information together into a distance metric, so we can find the k-nearest neighbors for Toy Story, for example.

I've rather arbitrarily computed this `ComputeDistance()` function, that takes two movie IDs and computes a distance score between the two. We're going to base this, first of all, on the similarity, using a cosine similarity metric, between the two genre vectors. Like I said, we're just going to take the list of genres for each movie and see how similar they are to each other. Again, a `0` indicates it's not part of that genre, a `1` indicates it is.

We will then compare the popularity scores and just take the raw difference, the absolute value of the difference between those two popularity scores and use that toward the distance metric as well. Then, we will use that information alone to define the distance between two movies. So, for example, if we compute the distance between movie IDs 2 and 4, this function would return some distance function based only on the popularity of that movie and on the genres of those movies.

Now, imagine a scatter plot if you will, like we saw back in our example from the previous sections, where one axis might be a measure of genre similarity, based on cosine metric, the other axis might be popularity, okay? We're just finding the distance between these two things:

```
from scipy import spatial

def ComputeDistance(a, b):
    genresA = a[1]
    genresB = b[1]
    genreDistance = spatial.distance.cosine(genresA, genresB)
    popularityA = a[2]
    popularityB = b[2]
    popularityDistance = abs(popularityA - popularityB)
    return genreDistance + popularityDistance
ComputeDistance(movieDict[2], movieDict[4])
```

For this example, where we're trying to compute the distance using our distance metric between movies 2 and 4, we end up with a score of 0.8:

```
0.8004574042309891
```

Remember, a far distance means it's not similar, right? We want the nearest neighbors, with the smallest distance. So, a score of 0.8 is a pretty high number on a scale of 0 to 1. So that's telling me that these movies really aren't similar. Let's do a quick sanity check and see what these movies really are:

```
print movieDict[2]
print movieDict[4]
```

It turns out it's the movies GoldenEye and Get Shorty, which are pretty darn different movies:

```
('GoldenEye (1995)', [0, 1, 1, 0, 0, 0, 0, 0, 0, 0, 0, 0, 0, 0, 0, 0, 1, 0, 0], 0.22298456260720412, 3.2061068702290076)
('Get Shorty (1995)', [0, 1, 0, 0, 0, 1, 0, 0, 1, 0, 0, 0, 0, 0, 0, 0, 0, 0, 0], 0.35677530017152659, 3.5502392344497609)
```

You know, you have James Bond action-adventure, and a comedy movie - not very similar at all! They're actually comparable in terms of popularity, but the genre difference did it in. Okay! So, let's put it all together!

Next, we're going to write a little bit of code to actually take some given movieID and find the KNN. So, all we have to do is compute the distance between Toy Story and all the other movies in our movie dictionary, and sort the results based on their distance score. That's what the following little snippet of code does. If you want to take a moment to wrap your head around it, it's fairly straightforward.

We have a little `getNeighbors()` function that will take the movie that we're interested in, and the K neighbors that we want to find. It'll iterate through every movie that we have; if it's actually a different movie than we're looking at, it will compute that distance score from before, append that to the list of results that we have, and sort that result. Then we will pluck off the K top results.

In this example, we're going to set *K* to 10, find the 10 nearest neighbors. We will find the 10 nearest neighbors using `getNeighbors()`, and then we will iterate through all these 10 nearest neighbors and compute the average rating from each neighbor. That average rating will inform us of our rating prediction for the movie in question.

As a side effect, we also get the 10 nearest neighbors based on our distance function, which we could call similar movies. So, that information itself is useful. Going back to that "Customers Who Watched Also Watched" example, if you wanted to do a similar feature that was just based on this distance metric and not actual behavior data, this might be a reasonable place to start, right?

```
import operator

def getNeighbors(movieID, K):
    distances = []
    for movie in movieDict:
        if (movie != movieID):
            dist = ComputeDistance(movieDict[movieID],
 movieDict[movie])
            distances.append((movie, dist))
    distances.sort(key=operator.itemgetter(1))
    neighbors = []
    for x in range(K):
        neighbors.append(distances[x][0])
    return neighbors

K = 10
avgRating = 0
neighbors = getNeighbors(1, K)
for neighbor in neighbors:
    avgRating += movieDict[neighbor][3]
    print movieDict[neighbor][0] + " " +
 str(movieDict[neighbor][3])
    avgRating /= float(K)
```

So, let's go ahead and run this, and see what we end up with. The output of the following code is as follows:

```
Liar Liar (1997) 3.15670103093
Aladdin (1992) 3.81278538813
Willy Wonka and the Chocolate Factory (1971) 3.63190184049
Monty Python and the Holy Grail (1974) 4.0664556962
Full Monty, The (1997) 3.92698412698
George of the Jungle (1997) 2.68518518519
Beavis and Butt-head Do America (1996) 2.78846153846
Birdcage, The (1996) 3.44368600683
Home Alone (1990) 3.08759124088
Aladdin and the King of Thieves (1996) 2.84615384615
```

The results aren't that unreasonable. So, we are using as an example the movie Toy Story, which is movieID 1, and what we came back with, for the top 10 nearest neighbors, are a pretty good selection of comedy and children's movies. So, given that Toy Story is a popular comedy and children's movie, we got a bunch of other popular comedy and children's movies; so, it seems to work! We didn't have to use a bunch of fancy collaborative filtering algorithms, these results aren't that bad.

Next, let's use KNN to predict the rating, where we're thinking of the rating as the classification in this example:

```
avgRating
```

Following is the output of the preceding code:

```
3.3445905900235564
```

We end up with a predicted rating of 3.34, which actually isn't all that different from the actual rating for that movie, which was 3.87. So not great, but it's not too bad either! I mean it actually works surprisingly well, given how simple this algorithm is!

Activity

Most of the complexity in this example was just in determining our distance metric, and you know we intentionally got a little bit fancy there just to keep it interesting, but you can do anything else you want to. So, if you want fiddle around with this, I definitely encourage you to do so. Our choice of 10 for K was completely out of thin air, I just made that up. How does this respond to different K values? Do you get better results with a higher value of K? Or with a lower value of K? Does it matter?

If you really want to do a more involved exercise you can actually try to apply it to train/test, to actually find the value of K that most optimally can predict the rating of the given movie based on KNN. And, you can use different distance metrics, I kind of made that up too! So, play around the distance metric, maybe you can use different sources of information, or weigh things differently. It might be an interesting thing to do. Maybe, popularity isn't really as important as the genre information, or maybe it's the other way around. See what impact that has on your results too. So, go ahead and mess with these algorithms, mess with the code and run with it, and see what you can get! And, if you do find a significant way of improving on this, share that with your classmates.

That is KNN in action! So, a very simple concept but it can be actually pretty powerful. So, there you have it: similar movies just based on the genre and popularity and nothing else. Works out surprisingly well! And, we used the concept of KNN to actually use those nearest neighbors to predict a rating for a new movie, and that actually worked out pretty well too. So, that's KNN in action, very simple technique but often it works out pretty darn good!

Dimensionality reduction and principal component analysis

Alright, time to get all trippy! We're going to talking about higher dimensions, and dimensionality reduction. Sounds scary! There is some fancy math involved, but conceptually it's not as hard to grasp as you might think. So, let's talk about dimensionality reduction and principal component analysis next. Very dramatic sounding! Usually when people talk about this, they're talking about a technique called principal component analysis or PCA, and a specific technique called singular value decomposition or SVD. So PCA and SVD are the topics of this section. Let's dive into it!

Dimensionality reduction

So, what is the curse of dimensionality? Well, a lot of problems can be thought of having many different dimensions. So, for example, when we were doing movie recommendations, we had attributes of various movies, and every individual movie could be thought of as its own dimension in that data space.

If you have a lot of movies, that's a lot of dimensions and you can't really wrap your head around more than 3, because that's what we grew up to evolve within. You might have some sort of data that has many different features that you care about. You know, in a moment we'll look at an example of flowers that we want to classify, and that classification is based on 4 different measurements of the flowers. Those 4 different features, those 4 different measurements can represent 4 dimensions, which again, is very hard to visualize.

For this reason, dimensionality reduction techniques exist to find a way to reduce higher dimensional information into lower dimensional information. Not only can that make it easier to look at, and classify things, but it can also be useful for things like compressing data. So, by preserving the maximum amount of variance, while we reduce the number of dimensions, we're more compactly representing a dataset. A very common application of dimensionality reduction is not just for visualization, but also for compression, and for feature extraction. We'll talk about that a little bit more in a moment.

A very simple example of dimensionality reduction can be thought of as k-means clustering:

So you know, for example, we might start off with many points that represent many different dimensions in a dataset. But, ultimately, we can boil that down to K different centroids, and your distance to those centroids. That's one way of boiling data down to a lower dimensional representation.

Principal component analysis

Usually, when people talk about dimensionality reduction, they're talking about a technique called principal component analysis. This is a much more-fancy technique, it gets into some pretty involved mathematics. But, at a high-level, all you need to know is that it takes a higher dimensional data space, and it finds planes within that data space and higher dimensions.

More Data Mining and Machine Learning Techniques

These higher dimensional planes are called hyper planes, and they are defined by things called eigenvectors. You take as many planes as you want dimensions in the end, project that data onto those hyperplanes, and those become the new axes in your lower dimensional data space:

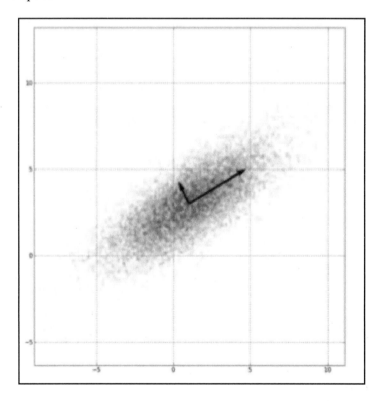

You know, unless you're familiar with higher dimensional math and you've thought about it before, it's going to be hard to wrap your head around! But, at the end of the day, it means we're choosing planes in a higher dimensional space that still preserve the most variance in our data, and project the data onto those higher dimensional planes that we then bring into a lower dimensional space, okay?

You don't really have to understand all the math to use it; the important point is that it's a very principled way of reducing a dataset down to a lower dimensional space while still preserving the variance within it. We talked about image compression as one application of this. So you know, if I want to reduce the dimensionality in an image, I could use PCA to boil it down to its essence.

Facial recognition is another example. So, if I have a dataset of faces, maybe each face represents a third dimension of 2D images, and I want to boil that down, SVD and principal component analysis can be a way to identify the features that really count in a face. So, it might end up focusing more on the eyes and the mouth, for example, those important features that are necessary for preserving the variance within that dataset. So, it can produce some very interesting and very useful results that just emerge naturally out of the data, which is kind of cool!

To make it real, we're going to use a simpler example, using what's called the Iris dataset. This is a dataset that's included with scikit-learn. It's used pretty commonly in examples, and here's the idea behind it: So, an Iris actually has 2 different kinds of petals on its flower. One's called a petal, which is the flower petals you're familiar with, and it also has something called a sepal, which is kind of this supportive lower set of petals on the flower.

We can take a bunch of different species of Iris, and measure the petal length and width, and the sepal length and width. So, together the length and width of the petal, and the length and width of the sepal are 4 different measurements that correspond to 4 different dimensions in our dataset. I want to use that to classify what species an Iris might belong to. Now, PCA will let us visualize this in 2 dimensions instead of 4, while still preserving the variance in that dataset. So, let's see how well that works and actually write some Python code to make PCA happen on the Iris dataset.

So, those were the concepts of dimensionality reduction, principal component analysis, and singular value decomposition. All big fancy words and yeah, it is kind of a fancy thing. You know, we're dealing with reducing higher dimensional spaces down to smaller dimensional spaces in a way that preserves their variance. Fortunately, scikit-learn makes this extremely easy to do, like 3 lines of code is all you need to actually apply PCA. So let's make that happen!

A PCA example with the Iris dataset

Let's apply principal component analysis to the Iris dataset. This is a 4D dataset that we're going to reduce down to 2 dimensions. We're going to see that we can actually still preserve most of the information in that dataset, even by throwing away half of the dimensions. It's pretty cool stuff, and it's pretty simple too. Let's dive in and do some principal component analysis and cure the curse of dimensionality. Go ahead and open up the `PCA.ipynb` file.

It's actually very easy to do using scikit-learn, as usual! Again, PCA is a dimensionality reduction technique. It sounds very science-fictiony, all this talk of higher dimensions. But, just to make it more concrete and real again, a common application is image compression. You can think of an image of a black and white picture, as 3 dimensions, where you have width, as your x-axis, and your y-axis of height, and each individual cell has some brightness value on a scale of 0 to 1, that is black or white, or some value in between. So, that would be 3D data; you have 2 spatial dimensions, and then a brightness and intensity dimension on top of that.

If you were to distill that down to say 2 dimensions alone, that would be a compressed image and, if you were to do that in a technique that preserved the variance in that image as well as possible, you could still reconstruct the image, without a whole lot of loss in theory. So, that's dimensionality reduction, distilled down to a practical example.

Now, we're going to use a different example here using the Iris dataset, and scikit-learn includes this. All it is is a dataset of various Iris flower measurements, and the species classification for each Iris in that dataset. And it has also, like I said before, the length and width measurement of both the petal and the sepal for each Iris specimen. So, between the length and width of the petal, and the length and width of the sepal we have 4 dimensions of feature data in our dataset.

We want to distill that down to something we can actually look at and understand, because your mind doesn't deal with 4 dimensions very well, but you can look at 2 dimensions on a piece of paper pretty easily. Let's go ahead and load that up:

```
from sklearn.datasets import load_iris
from sklearn.decomposition import PCA
import pylab as pl
from itertools import cycle

iris = load_iris()

numSamples, numFeatures = iris.data.shape
print numSamples
print numFeatures
print list(iris.target_names)
```

There's a handy dandy `load_iris()` function built into scikit-learn that will just load that up for you with no additional work; so you can just focus on the interesting part. Let's take a look at what that dataset looks like, the output of the preceding code is as follows:

```
150
4
['setosa', 'versicolor', 'virginica']
```

You can see that we are extracting the shape of that dataset, which means how many data points we have in it, that is 150, and how many features, or how many dimensions that dataset has, and that is 4. So, we have 150 Iris specimens in our dataset, with 4 dimensions of information. Again, that is the length and width of the sepal, and the length and width of the petal, for a total of 4 features, which we can think of as 4 dimensions.

We can also print out the list of target names in this dataset, which are the classifications, and we can see that each Iris belongs to one of three different species: Setosa, Versicolor, or Virginica. That's the data that we're working with: 150 Iris specimens, classified into one of 3 species, and we have 4 features associated with each Iris.

Let's look at how easy PCA is. Even though it's a very complicated technique under the hood, doing it is just a few lines of code. We're going to assign the entire Iris dataset and we're going to call it X. We will then create a PCA model, and we're going to keep `n_components=2`, because we want 2 dimensions, that is, we're going to go from 4 to 2.

We're going to use `whiten=True`, that means that we're going to normalize all the data, and make sure that everything is nice and comparable. Normally you will want to do that to get good results. Then, we will fit the PCA model to our Iris dataset X. We can use that model then, to transform that dataset down to 2 dimensions. Let's go ahead and run that. It happened pretty quickly!

```
X = iris.data
pca = PCA(n_components=2, whiten=True).fit(X)
X_pca = pca.transform(X)
```

Please think about what just happened there. We actually created a PCA model to reduce 4 dimensions down to 2, and it did that by choosing 2 4D vectors, to create hyperplanes around, to project that 4D data down to 2 dimensions. You can actually see what those 4D vectors are, those eigenvectors, by printing out the actual components of PCA. So, **PCA** stands for **Principal Component Analysis**, those principal components are the eigenvectors that we chose to define our planes about:

```
print pca.components_
```

Output to the preceding code is as follows:

```
[[ 0.36158968 -0.08226889  0.85657211  0.35884393]
 [-0.65653988 -0.72971237  0.1757674   0.07470647]]
```

You can actually look at those values, they're not going to mean a lot to you, because you can't really picture 4 dimensions anyway, but we did that just so you can see that it's actually doing something with principal components. So, let's evaluate our results:

```
print pca.explained_variance_ratio_
print sum(pca.explained_variance_ratio_)
```

The PCA model gives us back something called `explained_variance_ratio`. Basically, that tells you how much of the variance in the original 4D data was preserved as I reduced it down to 2 dimensions. So, let's go ahead and take a look at that:

```
[ 0.92461621  0.05301557]
0.977631775025
```

What it gives you back is actually a list of 2 items for the 2 dimensions that we preserved. This is telling me that in the first dimension I can actually preserve 92% of the variance in the data, and the second dimension only gave me an additional 5% of variance. If I sum it together, these 2 dimensions that I projected my data down into, I still preserved over 97% of the variance in the source data. We can see that 4 dimensions weren't really necessary to capture all the information in this dataset, which is pretty interesting. It's pretty cool stuff!

If you think about it, why do you think that might be? Well, maybe the overall size of the flower has some relationship to the species at its center. Maybe it's the ratio of length to width for the petal and the sepal. You know, some of these things probably move together in concert with each other for a given species, or for a given overall size of a flower. So, perhaps there are relationships between these 4 dimensions that PCA is extracting on its own. It's pretty cool, and pretty powerful stuff. Let's go ahead and visualize this.

The whole point of reducing this down to 2 dimensions was so that we could make a nice little 2D scatter plot of it, at least that's our objective for this little example here. So, we're going to do a little bit of Matplotlib magic here to do that. There is some sort of fancy stuff going on here that I should at least mention. So, what we're going to do is create a list of colors: red, green and blue. We're going to create a list of target IDs, so that the values 0, 1, and 2 map to the different Iris species that we have.

Chapter 7

What we're going to do is zip all this up with the actual names of each species. The for loop will iterate through the 3 different Iris species, and as it does that, we're going to have the index for that species, a color associated with it, and the actual human-readable name for that species. We'll take one species at a time and plot it on our scatter plot just for that species with a given color and the given label. We will then add in our legend and show the results:

```
colors = cycle('rgb')
target_ids = range(len(iris.target_names))
pl.figure()
for i, c, label in zip(target_ids, colors, iris.target_names):
    pl.scatter(X_pca[iris.target == i, 0], X_pca[iris.target == i, 1],
        c=c, label=label)
pl.legend()
pl.show()
```

The following is what we end up with:

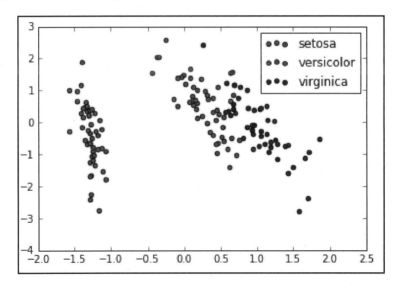

That is our 4D Iris data projected down to 2 dimensions. Pretty interesting stuff! You can see it still clusters together pretty nicely. You know, you have all the Virginicas sitting together, all the Versicolors sitting in the middle, and the Setosas way off on the left side. It's really hard to imagine what these actual values represent. But, the important point is, we've projected 4D data down to 2D, and in such a way that we still preserve the variance. We can still see clear delineations between these 3 species. A little bit of intermingling going on in there, it's not perfect you know. But by and large, it was pretty effective.

Activity

As you recall from `explained_variance_ratio`, we actually captured most of the variance in a single dimension. Maybe the overall size of the flower is all that really matters in classifying it; and you can specify that with one feature. So, go ahead and modify the results if you are feeling up to it. See if you can get away with 2 dimensions, or 1 dimension instead of 2! So, go change that `n_components` to `1`, and see what kind of variance ratio you get.

What happens? Does it makes sense? Play around with it, get some familiarity with it. That is dimensionality reduction, principal component analysis, and singular value decomposition all in action. Very, very fancy terms, and you know, to be fair it is some pretty fancy math under the hood. But as you can see, it's a very powerful technique and with scikit-learn, it's not hard to apply. So, keep that in your tool chest.

And there you have it! A 4D dataset of flower information boiled down to 2 dimensions that we can both easily visualize, and also still see clear delineations between the classifications that we're interested in. So, PCA works really well in this example. Again, it's a useful tool for things like compression, or feature extraction, or facial recognition as well. So, keep that in your toolbox.

Data warehousing overview

Next, we're going to talk a little bit about data warehousing. This is a field that's really been upended recently by the advent of Hadoop, and some big data techniques and cloud computing. So, a lot of big buzz words there, but concepts that are important for you to understand.

Let's dive in and explore these concepts! Let's talk about ELT and ETL, and data warehousing in general. This is more of a concept, as opposed to a specific practical technique, so we're going to talk about it conceptually. But, it is something that's likely to come up in the setting of a job interview. So, let's make sure you understand these concepts.

We'll start by talking about data warehousing in general. What is a data warehouse? Well, it's basically a giant database that contains information from many different sources and ties them together for you. For example, maybe you work at a big ecommerce company and they might have an ordering system that feeds information about the stuff people bought into your data warehouse.

You might also have information from web server logs that get ingested into the data warehouse. This would allow you to tie together browsing information on the website with what people ultimately ordered for example. Maybe you could also tie in information from your customer service systems, and measure if there's a relationship between browsing behavior and how happy the customers are at the end of the day.

A data warehouse has the challenge of taking data from many different sources, transforming them into some sort of schema that allows us to query these different data sources simultaneously, and it lets us make insights, through data analysis. So, large corporations and organizations have this sort of thing pretty commonly. We're going into the concept of big data here. You can have a giant Oracle database, for example, that contains all this stuff and maybe it's partitioned in some way, and replicated and it has all sorts of complexity. You can just query that through SQL, structured query language, or, through graphical tools, like Tableau which is a very popular one these days. That's what a data analyst does, they query large datasets using stuff like Tableau.

That's kind of the difference between a data analyst and a data scientist. You might be actually writing code to perform more advanced techniques on data that border on AI, as opposed to just using tools to extract graphs and relationships out of a data warehouse. It's a very complicated problem. At Amazon, we had an entire department for data warehousing that took care of this stuff full time, and they never had enough people, I can tell you that; it's a big job!

You know, there are a lot of challenges in doing data warehousing. One is data normalization: so, you have to figure out how do all the fields in these different data sources actually relate to each other? How do I actually make sure that a column in one data source is comparable to a column from another data source and has the same set of data, at the same scale, using the same terminology? How do I deal with missing data? How do I deal with corrupt data or data from outliers, or from robots and things like that? These are all very big challenges. Maintaining those data feeds is also a very big problem.

A lot can go wrong when you're importing all this information into your data warehouse, especially when you have a very large transformation that needs to happen to take the raw data, saved from web logs, into an actual structured database table that can be imported into your data warehouse. Scaling also can get tricky when you're dealing with a monolithic data warehouse. Eventually, your data will get so large that those transformations themselves start to become a problem. This starts to get into the whole topic of ELT versus ETL thing.

ETL versus ELT

Let's first talk about ETL. What does that stand for? It stands for extract, transform, and load - and that's sort of the conventional way of doing data warehousing.

Basically, first you extract the data that you want from the operational systems that you want. So, for example, I might extract all of the web logs from our web servers each day. Then I need to transform all that information into an actual structured database table that I can import into my data warehouse.

This transformation stage might go through every line of those web server logs, transform that into an actual table, where I'm plucking out from each web log line things like session ID, what page they looked at, what time it was, what the referrer was and things like that, and I can organize that into a tabular structure that I can then load into the data warehouse itself, as an actual table in a database. So, as data becomes larger and larger, that transformation step can become a real problem. Think about how much processing work is required to go through all of the web logs on Google, or Amazon, or any large website, and transform that into something a database can ingest. That itself becomes a scalability challenge and something that can introduce stability problems through the entire data warehouse pipeline.

That's where the concept of ELT comes in, and it kind of flips everything on its head. It says, "Well, what if we don't use a huge Oracle instance? What if instead we use some of these newer techniques that allow us to have a more distributed database over a Hadoop cluster that lets us take the power of these distributed databases like Hive, or Spark, or MapReduce, and use that to actually do the transformation after it's been loaded"

The idea here is we're going to extract the information we want as we did before, say from a set of web server logs. But then, we're going to load that straight in to our data repository, and we're going to use the power of the repository itself to actually do the transformation in place. So, the idea here is, instead of doing an offline process to transform my web logs, as an example, into a structured format, I'm just going to suck those in as raw text files and go through them one line at a time, using the power of something like Hadoop, to actually transform those into a more structured format that I can then query across my entire data warehouse solution.

Things like Hive let you host a massive database on a Hadoop cluster. There's things like Spark SQL that let you also do queries in a very SQL-like data warehouse-like manner, on a data warehouse that is actually distributed on Hadoop cluster. There are also distributed NoSQL data stores that can be queried using Spark and MapReduce. The idea is that instead of using a monolithic database for a data warehouse, you're instead using something built on top of Hadoop, or some sort of a cluster, that can actually not only scale up the processing and querying of that data, but also scale the transformation of that data as well.

Once again, you first extract your raw data, but then we're going to load it into the data warehouse system itself as is. And, then use the power of the data warehouse, which might be built on Hadoop, to do that transformation as the third step. Then I can query things together. So, it's a very big project, very big topic. You know, data warehousing is an entire discipline in and of itself. We're going to talk about Spark some more in this book very soon, which is one way of handling this thing - there's something called Spark SQL in particular that's relevant.

The overall concept here is that if you move from a monolithic database built on Oracle or MySQL to one of these more modern distributed databases built on top of Hadoop, you can take that transform stage and actually do that after you've loaded in the raw data, as opposed to before. That can end up being simpler and more scalable, and taking advantage of the power of large computing clusters that are available today.

That's ETL versus ELT, the legacy way of doing it with a lot of clusters all over the place in cloud-based computing versus a way that makes sense today, when we do have large clouds of computing available to us for transforming large datasets. That's the concept.

ETL is kind of the old school way of doing it, you transform a bunch of data offline before importing it in and loading it into a giant data warehouse, monolithic database. But with today's techniques, with cloud-based databases, and Hadoop, and Hive, and Spark, and MapReduce, you can actually do it a little bit more efficiently and take the power of a cluster to actually do that transformation step after you've loaded the raw data into your data warehouse.

This is really changing the field and it's important that you know about it. Again, there's a lot more to learn on the subject, so I encourage you to explore more on this topic. But, that's the basic concept, and now you know what people are talking about when they talk about ETL versus ELT.

Reinforcement learning

Our next topic's a fun one: reinforcement learning. We can actually use this idea with an example of Pac-Man. We can actually create a little intelligent Pac-Man agent that can play the game Pac-Man really well on its own. You'll be surprised how simple the technique is for building up the smarts behind this intelligent Pac-Man. Let's take a look!

So, the idea behind reinforcement learning is that you have some sort of agent, in this case Pac-Man, that explores some sort of space, and in our example that space will be the maze that Pac-Man is in. As it goes, it learns the value of different state changes within different conditions.

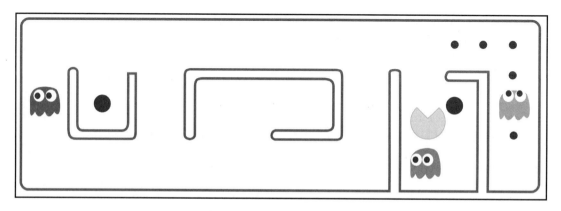

For example, in the preceding image, the state of Pac-Man might be defined by the fact that it has a ghost to the South, and a wall to the West, and empty spaces to the North and East, and that might define the current state of Pac-Man. The state changes it can take would be to move in a given direction. I can then learn the value of going in a certain direction. So, for example, if I were to move North, nothing would really happen, there's no real reward associated with that. But, if I were to move South I would be destroyed by the ghost, and that would be a negative value.

As I go and explore the entire space, I can build up a set of all the possible states that Pac-Man can be in, and the values associated with moving in a given direction in each one of those states, and that's reinforcement learning. And as it explores the whole space, it refines these reward values for a given state, and it can then use those stored reward values to choose the best decision to make given a current set of conditions. In addition to Pac-Man, there's also a game called Cat & Mouse that is an example that's used commonly that we'll look at later.

The benefit of this technique is that once you've explored the entire set of possible states that your agent can be in, you can very quickly have a very good performance when you run different iterations of this. So, you know, you can basically make an intelligent Pac-Man by running reinforcement learning and letting it explore the values of different decisions it can make in different states and then storing that information, to very quickly make the right decision given a future state that it sees in an unknown set of conditions.

Q-learning

So, a very specific implementation of reinforcement learning is called Q-learning, and this formalizes what we just talked about a little bit more:

- So again, you start with a set of environmental states of the agent (Is there a ghost next to me? Is there a power pill in front of me? Things like that.), we're going to call that s.
- I have a set of possible actions that I can take in those states, we're going to call that set of actions a. In the case of Pac-Man, those possible actions are move up, down, left, or right.
- Then we have a value for each state/action pair that we'll call Q; that's why we call it Q-learning. So, for each state, a given set of conditions surrounding Pac-Man, a given action will have a value Q. So, moving up might have a given value Q, moving down might have a negative Q value if it means encountering a ghost, for example.

So, we start off with a Q value of 0 for every possible state that Pac-Man could be in. And, as Pac-Man explores a maze, as bad things happen to Pac-Man, we reduce the Q value for the state that Pac-Man was in at the time. So, if Pac-Man ends up getting eaten by a ghost, we penalize whatever he did in that current state. As good things happen to Pac-Man, as he eats a power pill, or eats a ghost, we'll increase the Q value for that action, for the state that he was in. Then, what we can do is use those Q values to inform Pac-Man's future choices, and sort of build a little intelligent agent that can perform optimally, and make a perfect little Pac-Man. From the same image of Pac-Man that we saw just above, we can further define the current state of Pac-Man by defining that he has a wall to the West, empty space to the North and East, a ghost to the South.

We can look at the actions he can take: he can't actually move left at all, but he can move up, down, or right, and we can assign a value to all those actions. By going up or right, nothing really happens at all, there's no power pill or dots to consume. But if he goes left, that's definitely a negative value. We can say for the state given by the current conditions that Pac-Man is surrounded by, moving down would be a really bad choice; there should be a negative Q value for that. Moving left just can't be done at all. Moving up or right or staying neutral, the Q value would remain 0 for those action choices for that given state.

Now, you can also look ahead a little bit, to make an even more intelligent agent. So, I'm actually two steps away from getting a power pill here. So, as Pac-Man were to explore this state, if I were to hit the case of eating that power pill on the next state, I could actually factor that into the Q value for the previous state. If you just have some sort of a discount factor, based on how far away you are in time, how many steps away you are, you can factor that all in together. So, that's a way of actually building in a little bit of memory into the system. You can "look ahead" more than one step by using a discount factor when computing Q (here s is previous state, s' is current state):

$$Q(s,a) \mathrel{+}= discount * (reward(s,a) + max(Q(s')) - Q(s,a))$$

So, the Q value that I experience when I consume that power pill might actually give a boost to the previous Q values that I encountered along the way. So, that's a way to make Q-learning even better.

The exploration problem

One problem that we have in reinforcement learning is the exploration problem. How do I make sure that I efficiently cover all the different states and actions within those states during the exploration phase?

The simple approach

One simple approach is to always choose the action for a given state with the highest Q value that I've computed so far, and if there's a tie, just choose at random. So, initially all of my Q values might be 0, and I'll just pick actions at random at first.

As I start to gain information about better Q values for given actions and given states, I'll start to use those as I go. But, that ends up being pretty inefficient, and I can actually miss a lot of paths that way if I just tie myself into this rigid algorithm of always choosing the best Q value that I've computed thus far.

The better way

So, a better way is to introduce a little bit of random variation into my actions as I'm exploring. So, we call that an epsilon term. So, suppose we have some value, that I roll the dice, I have a random number. If it ends up being less than this epsilon value, I don't actually follow the highest Q value; I don't do the thing that makes sense, I just take a path at random to try it out, and see what happens. That actually lets me explore a much wider range of possibilities, a much wider range of actions, for a wider range of states more efficiently during that exploration stage.

So, what we just did can be described in very fancy mathematical terms, but you know conceptually it's pretty simple.

Fancy words

I explore some set of actions that I can take for a given set of states, I use that to inform the rewards associated with a given action for a given set of states, and after that exploration is done I can use that information, those Q values, to intelligently navigate through an entirely new maze for example.

This can also be called a Markov decision process. So again, a lot of data science is just assigning fancy, intimidating names to simple concepts, and there's a ton of that in reinforcement learning.

Markov decision process

So, if you look up the definition of Markov decision processes, it is "a mathematical framework for modeling decision making in situations where outcomes are partly random and partly under the control of a decision maker".

- **Decision making**: What action do we take given a set of possibilities for a given state?
- **In situations where outcomes are partly random**: Hmm, kind of like our random exploration there.
- **Partly under the control of a decision maker**: The decision maker is our Q values that we computed.

So, MDPs, Markov decision processes, are a fancy way of describing our exploration algorithm that we just described for reinforcement learning. The notation is even similar, states are still described as s, and s' is the next state that we encounter. We have state transition functions that are defined as P_a for a given state of s and s'. We have our Q values, which are basically represented as a reward function, an R_a value for a given s and s'. So, moving from one state to another has a given reward associated with it, and moving from one state to another is defined by a state transition function:

- States are still described as *s* and *s''*
- State transition functions are described as *Pa(s,s')*
- Our Q values are described as a reward function *Ra(s,s')*

So again, describing what we just did, only a mathematical notation, and a fancier sounding word, Markov decision processes. And, if you want to sound even smarter, you can also call a Markov decision process by another name: a discrete time stochastic control process. That sounds intelligent! But the concept itself is the same thing that we just described.

Dynamic programming

So, even more fancy words: dynamic programming can be used to describe what we just did as well. Wow! That sounds like artificial intelligence, computers programming themselves, Terminator 2, Skynet stuff. But no, it's just what we just did. If you look up the definition of dynamic programming, it is a method for solving a complex problem by breaking it down into a collection of simpler subproblems, solving each of those subproblems just once, and storing their solutions ideally, using a memory-based data structure.

The next time the same subproblem occurs, instead of recomputing its solution, one simply looks up the previously computed solution thereby saving computation time at the expense of a (hopefully) modest expenditure in storage space:

- **A method for solving a complex problem**: Same as creating an intelligent Pac-Man, that's a pretty complicated end result.
- **By breaking it down into a collection of simpler subproblems**: So, for example, what is the optimal action to take for a given state that Pac-Man might be in. There are many different states Pac-Man could find himself in, but each one of those states represents a simpler subproblem, where there's a limited set of choices I can make, and there's one right answer for the best move to make.

- **Storing their solutions**: Those solutions being the *Q* values that I associated with each possible action at each state.
- **Ideally, using a memory-based data structure**: Well, of course I need to store those *Q* values and associate them with states somehow, right?
- **The next time the same subproblem occurs**: The next time Pac-Man is in a given state that I have a set of *Q* values for.
- **Instead of recomputing its solution, one simply looks up the previously computed solution**: The *Q* value I already have from the exploration stage.
- **Thereby saving computation time at the expense of a modest expenditure in storage space**: That's exactly what we just did with reinforcement learning.

We have a complicated exploration phase that finds the optimal rewards associated with each action for a given state. Once we have that table of the right action to take for a given state, we can very quickly use that to make our Pac-Man move in an optimal manner in a whole new maze that he hasn't seen before. So, reinforcement learning is also a form of dynamic programming. Wow!

To recap, you can make an intelligent Pac-Man agent by just having it semi-randomly explore different choices of movement given different conditions, where those choices are actions and those conditions are states. We keep track of the reward or penalty associated with each action or state as we go, and we can actually discount, going back multiple steps if you want to make it even better.

Then we store those *Q* values that we end up associating with each state, and we can use that to inform its future choices. So we can go into a whole new maze, and have a really smart Pac-Man that can avoid the ghosts and eat them up pretty effectively, all on its own. It's a pretty simple concept, very powerful though. You can also say that you understand a bunch of fancy terms because it's all called the same thing. Q-learning, reinforcement learning, Markov decision processes, dynamic programming: all tied up in the same concept.

I don't know, I think it's pretty cool that you can actually make sort of an artificially intelligent Pac-Man through such a simple technique, and it really does work! If you want to go look at it in more detail, following are a few examples you can look at that have one actual source code you can look at, and potentially play with, **Python Markov Decision Process Toolbox**: http://pymdptoolbox.readthedocs.org/en/latest/api/mdp.html.

There is a Python Markov decision process toolbox that wraps it up in all that terminology we talked about. There's an example you can look at, a working example of the cat and mouse game, which is similar. And, there is actually a Pac-Man example you can look at online as well, that ties in more directly with what we were talking about. Feel free to explore these links, and learn even more about it.

And so that's reinforcement learning. More generally, it's a useful technique for building an agent that can navigate its way through a possible different set of states that have a set of actions that can be associated with each state. So, we've talked about it mostly in the context of a maze game. But, you can think more broadly, and you know whenever you have a situation where you need to predict behavior of something given a set of current conditions and a set of actions it can take. Reinforcement learning and Q-learning might be a way of doing it. So, keep that in mind!

Summary

In this chapter, we saw one of the simplest techniques of machine learning called k-nearest neighbors. We also looked at an example of KNN which predicts the rating for a movie. We analysed the concepts of dimensionality reduction and principal component analysis and saw an example of PCA, which reduced 4D data to two dimensions while still preserving its variance.

Next, we learned the concept of data warehousing and saw how using the ELT process instead of ETL makes more sense today. We walked through the concept of reinforcement learning and saw how it is used behind the Pac-Man game. Finally, we saw some fancy words used for reinforcement learning (Q-learning, Markov decision process, and dynamic learning). In the next chapter, we'll see how to deal with real-world data.

8
Dealing with Real-World Data

In this chapter, we're going to talk about the challenges of dealing with real-world data, and some of the quirks you might run into. The chapter starts by talking about the bias-variance trade-off, which is kind of a more principled way of talking about the different ways you might overfit and underfit data, and how it all interrelates with each other. We then talk about the k-fold cross-validation technique, which is an important tool in your chest to combat overfitting, and look at how to implement it using Python.

Next, we analyze the importance of cleaning your data and normalizing it before actually applying any algorithms on it. We see an example to determine the most popular pages on a website which will demonstrate the importance of cleaning data. The chapter also covers the importance of remembering to normalize numerical data. Finally, we look at how to detect outliers and deal with them.

Specifically, this chapter covers the following topics:

- Analyzing the bias/variance trade-off
- The concept of k-fold cross-validation and its implementation
- The importance of cleaning and normalizing data
- An example to determine the popular pages of a website
- Normalizing numerical data
- Detecting outliers and dealing with them

Bias/variance trade-off

One of the basic challenges that we face when dealing with real-world data is overfitting versus underfitting your regressions to that data, or your models, or your predictions. When we talk about underfitting and overfitting, we can often talk about that in the context of bias and variance, and the bias-variance trade-off. So, let's talk about what that means.

So conceptually, bias and variance are pretty simple. Bias is just how far off you are from the correct values, that is, how good are your predictions overall in predicting the right overall value. If you take the mean of all your predictions, are they more or less on the right spot? Or are your errors all consistently skewed in one direction or another? If so, then your predictions are biased in a certain direction.

Variance is just a measure of how spread out, how scattered your predictions are. So, if your predictions are all over the place, then that's high variance. But, if they're very tightly focused on what the correct values are, or even an incorrect value in the case of high bias, then your variance is small.

Let's look at some examples. Let's imagine that the following dartboard represents a bunch of predictions we're making where the real value we're trying to predict is in the center of the bullseye:

- Starting with the dartboard in the upper left-hand corner, you can see that our points are all scattered about the center. So overall, you know the mean error comes out to be pretty close to reality. Our bias is actually very low, because our predictions are all around the same correct point. However, we have very high variance, because these points are scattered about all over the place. So, this is an example of low bias and high variance.

- If we move on to the dartboard in the upper right corner, we see that our points are all consistently skewed from where they should be, to the Northwest. So this is an example of high bias in our predictions, where they're consistently off by a certain amount. We have low variance because they're all clustered tightly around the wrong spot, but at least they're close together, so we're being consistent in our predictions. That's low variance. But, the bias is high. So again, this is high bias, low variance.
- In the dartboard in the lower left corner, you can see that our predictions are scattered around the wrong mean point. So, we have high bias; everything is skewed to some place where it shouldn't be. But our variance is also high. So, this is kind of the worst of both worlds here; we have high bias and high variance in this example.
- Finally, in a wonderful perfect world, you would have an example like the lower right dartboard, where we have low bias, where everything is centered around where it should be, and low variance, where things are all clustered pretty tightly around where they should be. So, in a perfect world that's what you end up with.

In reality, you often need to choose between bias and variance. It comes down to over fitting Vs underfitting your data. Let's take a look at the following example:

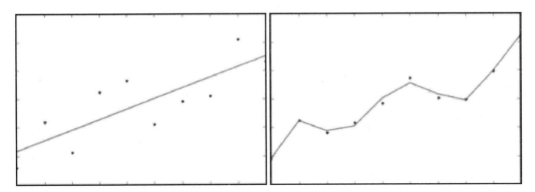

It's a little bit of a different way of thinking of bias and variance. So, in the left graph, we have a straight line, and you can think of that as having very low variance, relative to these observations. So, there's not a lot of variance in this line, that is, there is low variance. But the bias, the error from each individual point, is actually high.

Dealing with Real-World Data

Now, contrast that to the overfitted data in the graph at the right, where we've kind of gone out of our way to fit the observations. The line has high variance, but low bias, because each individual point is pretty close to where it should be. So, this is an example of where we traded off variance for bias.

At the end of the day, you're not out to just reduce bias or just reduce variance, you want to reduce error. That's what really matters, and it turns out you can express error as a function of bias and variance:

$$Error = Bias^2 + Variance$$

Looking at this, error is equal to bias squared plus variance. So, these things both contribute to the overall error, with bias actually contributing more. But keep in mind, it's error you really want to minimize, not the bias or the variance specifically, and that an overly complex model will probably end up having a high variance and low bias, whereas a too simple model will have low variance and high bias. However, they could both end up having similar error terms at the end of the day. You just have to find the right happy medium of these two things when you're trying to fit your data. We'll talk about some more principled ways of actually avoiding overfitting in our forthcoming sections. But, it's just the concept of bias and variance that I want to get across, because people do talk about it and you're going to be expected to know what means.

Now let's tie that back to some earlier concepts in this book. For example, in k-nearest neighbors if we increase the value of K, we start to spread out our neighborhood that were averaging across to a larger area. That has the effect of decreasing variance because we're kind of smoothing things out over a larger space, but it might increase our bias because we'll be picking up a larger population that may be less and less relevant to the point we started from. By smoothing out KNN over a larger number of neighbors, we can decrease the variance because we're smoothing things out over more values. But, we might be introducing bias because we're introducing more and more points that are less than less related to the point we started with.

Decision trees is another example. We know that a single decision tree is prone to overfitting, so that might imply that it has a high variance. But, random forests seek to trade off some of that variance for bias reduction, and it does that by having multiple trees that are randomly variant and averages all their solutions together. It's like when we average things out by increasing K in KNN: we can average out the results of a decision tree by using more than one decision tree using random forests similar idea.

This is bias-variance trade-off. You know the decision you have to make between how overall accurate your values are, and how spread out they are or how tightly clustered they are. That's the bias-variance trade-off and they both contribute to the overall error, which is the thing you really care about minimizing. So, keep those terms in mind!

K-fold cross-validation to avoid overfitting

Earlier in the book, we talked about train and test as a good way of preventing overfitting and actually measuring how well your model can perform on data it's never seen before. We can take that to the next level with a technique called k-fold cross-validation. So, let's talk about this powerful tool in your arsenal for fighting overfitting; k-fold cross-validation and learn how that works.

To recall from train/test, the idea was that we split all of our data that we're building a machine learning model based off of into two segments: a training dataset, and a test dataset. The idea is that we train our model only using the data in our training dataset, and then we evaluate its performance using the data that we reserved for our test dataset. That prevents us from overfitting to the data that we have because we're testing the model against data that it's never seen before.

However, train/test still has its limitations: you could still end up overfitting to your specific train/test split. Maybe your training dataset isn't really representative of the entire dataset, and too much stuff ended up in your training dataset that skews things. So, that's where k-fold cross-validation comes in, it takes train/test and kicks it up a notch.

The idea, although it sounds complicated, is fairly simple:

1. Instead of dividing our data into two buckets, one for training and one for testing, we divide it into K buckets.
2. We reserve one of those buckets for testing purposes, for evaluating the results of our model.
3. We train our model against the remaining buckets that we have, K-1, and then we take our test dataset and use that to evaluate how well our model did amongst all of those different training datasets.
4. We average those resulting error metrics, that is, those r-squared values, together to get a final error metric from k-fold cross-validation.

That's all it is. It is a more robust way of doing train/test, and that's one way of doing it.

Now, you might think to yourself well, what if I'm overfitting to that one test dataset that I reserved? I'm still using the same test dataset for every one of those training datasets. What if that test dataset isn't really representative of things either?

There are variations of k-fold cross-validation that will randomize that as well. So, you could randomly pick what the training dataset is as well each time around, and just keep randomly assigning things to different buckets and measuring the results. But usually, when people talk about k-fold cross-validation, they're talking about this specific technique where you reserve one bucket for testing, and the remaining buckets for training, and you evaluate all of your training datasets against the test dataset when you build a model for each one.

Example of k-fold cross-validation using scikit-learn

Fortunately, scikit-learn makes this really easy to do, and it's even easier than doing normal train/test! It's extremely simple to do k-fold cross-validation, so you may as well just do it.

Now, the way this all works in practice is you will have a model that you're trying to tune, and you will have different variations of that model, different parameters you might want to tweak on it, right?

Like, for example, the degree of polynomial for a polynomial fit. So, the idea is to try different values of your model, different variations, measure them all using k-fold cross-validation, and find the one that minimizes error against your test dataset. That's kind of your sweet spot there. In practice, you want to use k-fold cross-validation to measure the accuracy of your model against a test dataset, and just keep refining that model, keep trying different values within it, keep trying different variations of that model or maybe even different models entirely, until you find the technique that reduces error the most, using k-fold cross validation.

Let's go dive into an example and see how it works. We're going to apply this to our Iris dataset again, revisiting SVC, and we'll play with k-fold cross-validation and see how simple it is. Let's actually put k-fold cross-validation and train/test into practice here using some real Python code. You'll see it's actually very easy to use, which is a good thing because this is a technique you should be using to measure the accuracy, the effectiveness of your models in supervised learning.

Please go ahead and open up the `KFoldCrossValidation.ipynb` and follow along if you will. We're going to look at the Iris dataset again; remember we introduced this when we talk about dimensionality reduction?

Just to refresh your memory, the Iris dataset contains a set of 150 Iris flower measurements, where each flower has a length and width of its petal, and a length and width of its sepal. We also know which one of 3 different species of Iris each flower belongs to. The challenge here is to create a model that can successfully predict the species of an Iris flower, just given the length and width of its petal and sepal. So, let's go ahead and do that.

We're going to use the SVC model. If you remember back again, that's just a way of classifying data that's pretty robust. There's a section on that if you need to go and refresh your memory:

```
import numpy as np
from sklearn import cross_validation
from sklearn import datasets
from sklearn import svm

iris = datasets.load_iris()

# Split the iris data into train/test data sets with
#40% reserved for testing
X_train, X_test, y_train, y_test =
cross_validation.train_test_split(iris.data,
                                  iris.target, test_size=0.4,
random_state=0)

# Build an SVC model for predicting iris classifications
#using training data
clf = svm.SVC(kernel='linear', C=1).fit(X_train, y_train)

# Now measure its performance with the test data
clf.score(X_test, y_test)
```

What we do is use the `cross_validation` library from scikit-learn, and we start by just doing a conventional train test split, just a single train/test split, and see how that will work.

To do that we have a `train_test_split()` function that makes it pretty easy. So, the way this works is we feed into `train_test_split()` a set of feature data. `iris.data` just contains all the actual measurements of each flower. `iris.target` is basically the thing we're trying to predict.

In this case, it contains all the species for each flower. `test_size` says what percentage do we want to train versus test. So, 0.4 means we're going to extract 40% of that data randomly for testing purposes, and use 60% for training purposes. What this gives us back is 4 datasets, basically, a training dataset and a test dataset for both the feature data and the target data. So, `X_train` ends up containing 60% of our Iris measurements, and `X_test` contains 40% of the measurements used for testing the results of our model. `y_train` and `y_test` contain the actual species for each one of those segments.

Then after that we go ahead and build an SVC model for predicting Iris species given their measurements, and we build that only using the training data. We fit this SVC model, using a linear kernel, using only the training feature data, and the training species data, that is, target data. We call that model `clf`. Then, we call the `score()` function on `clf` to just measure its performance against our test dataset. So, we score this model against the test data we reserved for the Iris measurements, and the test Iris species, and see how well it does:

```
Out[2]:  0.96666666666666667
```

It turns out it does really well! Over 96% of the time, our model is able to correctly predict the species of an Iris that it had never seen before, just based on the measurements of that Iris. So that's pretty cool!

But, this is a fairly small dataset, about 150 flowers if I remember right. So, we're only using 60% of 150 flowers for training and only 40% of 150 flowers for testing. These are still fairly small numbers, so we could still be overfitting to our specific train/test split that we made. So, let's use k-fold cross-validation to protect against that. It turns out that using k-fold cross-validation, even though it's a more robust technique, is actually even easier to use than train/test. So, that's pretty cool! So, let's see how that works:

```
# We give cross_val_score a model, the entire data set and its "real"
values, and the number of folds:
scores = cross_validation.cross_val_score(clf, iris.data, iris.target,
cv=5)

# Print the accuracy for each fold:
print scores

# And the mean accuracy of all 5 folds:
print scores.mean()
```

We have a model already, the SVC model that we defined for this prediction, and all you need to do is call `cross_val_score()` on the `cross_validation` package. So, you pass in this function a model of a given type (`clf`), the entire dataset that you have of all of the measurements, that is, all of my feature data (`iris.data`) and all of my target data (all of the species), `iris.target`.

I want `cv=5` which means it's actually going to use 5 different training datasets while reserving 1 for testing. Basically, it's going to run it 5 times, and that's all we need to do. That will automatically evaluate our model against the entire dataset, split up five different ways, and give us back the individual results.

If we print back the output of that, it gives us back a list of the actual error metric from each one of those iterations, that is, each one of those folds. We can average those together to get an overall error metric based on k-fold cross-validation:

```
[ 1.          1.          0.9         0.93333333  1.         ]
0.966666666667
```

When we do this over 5 folds, we can see that our results are even better than we thought! 98% accuracy. So that's pretty cool! In fact, in a couple of the runs we had perfect accuracy. So that's pretty amazing stuff.

Now let's see if we can do even better. We used a linear kernel before, what if we used a polynomial kernel and got even fancier? Will that be overfitting or will it actually better fit the data that we have? That kind of depends on whether there's actually a linear relationship or polynomial relationship between these petal measurements and the actual species or not. So, let's try that out:

```
clf = svm.SVC(kernel='poly', C=1).fit(X_train, y_train)
scores = cross_validation.cross_val_score(clf, iris.data, iris.target, cv=5)
print scores
print scores.mean()
```

We'll just run this all again, using the same technique. But this time, we're using a polynomial kernel. We'll fit that to our training dataset, and it doesn't really matter where you fit to in this case, because `cross_val_score()` will just keep re-running it for you:

```
[ 1.          1.          0.9         0.93333333  1.         ]
0.966666666667
```

Dealing with Real-World Data

It turns out that when we use a polynomial fit, we end up with an overall score that's even lower than our original run. So, this tells us that the polynomial kernel is probably overfitting. When we use k-fold cross-validation it reveals an actual lower score than with our linear kernel.

The important point here is that if we had just used a single train/test split, we wouldn't have realized that we were overfitting. We would have actually gotten the same result if we just did a single train/test split here as we did on the linear kernel. So, we might inadvertently be overfitting our data there, and not have even known it had we not use k-fold cross-validation. So, this is a good example of where k-fold comes to the rescue, and warns you of overfitting, where a single train/test split might not have caught that. So, keep that in your tool chest.

If you want to play around with this some more, go ahead and try different degrees. So, you can actually specify a different number of degrees. The default is 3 degrees for the polynomial kernel, but you can try a different one, you can try two.

Does that do better? If you go down to one, that degrades basically to a linear kernel, right? So, maybe there is still a polynomial relationship and maybe it's only a second degree polynomial. Try it out and see what you get back. That's k-fold cross-validation. As you can see, it's very easy to use thanks to scikit-learn. It's an important way to measure how good your model is in a very robust manner.

Data cleaning and normalisation

Now, this is one of the simplest, but yet it might be the most important section in this whole book. We're going to talk about cleaning your input data, which you're going to spend a lot of your time doing.

How well you clean your input data and understand your raw input data is going to have a huge impact on the quality of your results - maybe even more so than what model you choose or how well you tune your models. So, pay attention; this is important stuff!

 Cleaning your raw input data is often the most important, and time-consuming, part of your job as a data scientist!

Let's talk about an inconvenient truth of data science, and that's that you spend most of your time actually just cleaning and preparing your data, and actually relatively little of it analyzing it and trying out new algorithms. It's not quite as glamorous as people might make it out to be all the time. But, this is an extremely important thing to pay attention to.

There are a lot of different things that you might find in raw data. Data that comes in to you, just raw data, is going to be very dirty, it's going to be polluted in many different ways. If you don't deal with it it's going to skew your results, and it will ultimately end up in your business making the wrong decisions.

If it comes back that you made a mistake where you ingested a bunch of bad data and didn't account for it, didn't clean that data up, and what you told your business was to do something based on those results that later turn out to be completely wrong, you're going to be in a lot of trouble! So, pay attention!

There are a lot of different kinds of problems and data that you need to watch out for:

- **Outliers**: So maybe you have people that are behaving kind of strangely in your data, and when you dig into them, they turn out to be data you shouldn't be looking at the in first place. A good example would be if you're looking at web log data, and you see one session ID that keeps coming back over, and over, and over again, and it keeps doing something at a ridiculous rate that a human could never do. What you're probably seeing there is a robot, a script that's being run somewhere to actually scrape your website. It might even be some sort of malicious attack. But at any rate, you don't want that behavior data informing your models that are meant to predict the behavior of real human beings using your website. So, watching for outliers is one way to identify types of data that you might want to strip out of your model when you're building it.
- **Missing data**: What do you do when data's just not there? Going back to the example of a web log, you might have a referrer in that line, or you might not. What do you do if it's not there? Do you create a new classification for missing, or not specified? Or do you throw that line out entirely? You have to think about what the right thing to do is there.
- **Malicious data**: There might be people trying to game your system, there might be people trying to cheat the system, and you don't want those people getting away with it. Let's say you're making a recommender system. There could be people out there trying to fabricate behavior data in order to promote their new item, right? So, you need to be on the lookout for that sort of thing, and make sure that you're identifying the shilling attacks, or other types of attacks on your input data, and filtering them out from results and don't let them win.

- **Erroneous data**: What if there's a software bug somewhere in some system that's just writing out the wrong values in some set of situations? It can happen. Unfortunately, there's no good way for you to know about that. But, if you see data that just looks fishy or the results don't make sense to you, digging in deeply enough can sometimes uncover an underlying bug that's causing the wrong data to be written in the first place. Maybe things aren't being combined properly at some point. Maybe sessions aren't being held throughout the entire session. People might be dropping their session ID and getting new session IDs as they go through a website, for example.
- **Irrelevant data**: A very simple one here. Maybe you're only interested in data from New York City people, or something for some reason. In that case all the data from people from the rest of the world is irrelevant to what you're trying to find out. The first thing you want to do is just throw all that data that away and restrict your data, whittle it down to the data that you actually care about.
- **Inconsistent data**: This is a huge problem. For example, in addresses, people can write the same address in many different ways: they might abbreviate street or they might not abbreviate street, they might not put street at the end of the street name at all. They might combine lines together in different ways, they might spell things differently, they might use a zip code in the US or zip plus 4 code in the US, they might have a country on it, they might not have a country on it. You need to somehow figure out what are the variations that you see and how can you normalize them all together.
- Maybe I'm looking at data about movies. A movie might have different names in different countries, or a book might have different names in different countries, but they mean the same thing. So, you need to look out for these things where you need to normalize your data, where the same data can be represented in many different ways, and you need to combine them together in order to get the correct results.
- **Formatting**: This can also be an issue; things can be inconsistently formatted. Take the example of dates: in the US we always do month, day, year (MM/DD/YY), but in other countries they might do day, month, year (DD/MM/YY), who knows. You need to be aware of these formatting differences. Maybe phone numbers have parentheses around the area code, maybe they don't; maybe they have dashes between each section of the numbers, maybe they don't; maybe social security numbers have dashes, maybe they don't. These are all things that you need to watch out for, and you need to make sure that variations in formatting don't get treated as different entities, or different classifications during your processing.

So, there are lots of things to watch out for, and the previous list names just the main ones to be aware of. Remember: garbage in, garbage out. Your model is only as good as the data that you give to it, and this is extremely, extremely true! You can have a very simple model that performs very well if you give it a large amount of clean data, and it could actually outperform a complex model on a more dirty dataset.

Therefore, making sure that you have enough data, and high-quality data is often most of the battle. You'd be surprised how simple some of the most successful algorithms used in the real world are. They're only successful by virtue of the quality of the data going into it, and the amount of data going into it. You don't always need fancy techniques to get good results. Often, the quality and quantity of your data counts just as much as anything else.

Always question your results! You don't want to go back and look for anomalies in your input data only when you get a result that you don't like. That will introduce an unintentional bias into your results where you're letting results that you like, or expect, go through unquestioned, right? You want to question things all the time to make sure that you're always looking out for these things because even if you find a result you like, if it turns out to be wrong, it's still wrong, it's still going to be informing your company in the wrong direction. That could come back to bite you later on.

As an example, I have a website called No-Hate News. It's non-profit, so I'm not trying to make any money by telling you about it. Let's say I just want to find the most popular pages on this website that I own. That sounds like a pretty simple problem, doesn't it? I should just be able to go through my web logs, and count up how many hits each page has, and sort them, right? How hard can it be?! Well, turns out it's really hard! So, let's dive into this example and see why it's difficult, and see some examples of real-world data cleanup that has to happen.

Cleaning web log data

We're going to show the importance of cleaning your data. I have some web log data from a little website that I own. We are just going to try to find the top viewed pages on that website. Sounds pretty simple, but as you'll see, it's actually quite challenging! So, if you want to follow along, the `TopPages.ipynb` is the notebook that we're working from here. Let's start!

Dealing with Real-World Data

I actually have an access log that I took from my actual website. It's a real HTTP access log from Apache and is included in your book materials. So, if you do want to play along here, make sure you update the path to move the access log to wherever you saved the book materials:

```
logPath = "E:\\sundog-consult\\Packt\\DataScience\\access_log.txt"
```

Applying a regular expression on the web log

So, I went and got the following little snippet of code off of the Internet that will parse an Apache access log line into a bunch of fields:

```
format_pat= re.compile(
    r"(?P<host>[\d\.]+)\s"
    r"(?P<identity>\S*)\s"
    r"(?P<user>\S*)\s"
    r"\[(?P<time>.*?)\]\s"
    r'"(?P<request>.*?)"\s'
    r"(?P<status>\d+)\s"
    r"(?P<bytes>\S*)\s"
    r'"(?P<referer>.*?)"\s'
    r'"(?P<user_agent>.*?)"\s*'
)
```

This code contains things like the host, the user, the time, the actual page request, the status, the referrer, `user_agent` (meaning which browser actually was used to view this page). It builds up what's called a regular expression, and we're using the `re` library to use it. That's basically a very powerful language for doing pattern matching on a large string. So, we can actually apply this regular expression to each line of our access log, and automatically group the bits of information in that access log line into these different fields. Let's go ahead and run this.

The obvious thing to do here, let's just whip up a little script that counts up each URL that we encounter that was requested, and keeps count of how many times it was requested. Then we can sort that list and get our top pages, right? Sounds simple enough!

So, we're going to construct a little Python dictionary called `URLCounts`. We're going to open up our log file, and for each line, we're going to apply our regular expression. If it actually comes back with a successful match for the pattern that we're trying to match, we'll say, Okay this looks like a decent line in our access log.

Let's extract the request field out of it, which is the actual HTTP request, the page which is actually being requested by the browser. We're going to split that up into its three components: it consists of an action, like get or post; the actual URL being requested; and the protocol being used. Given that information split out, we can then just see if that URL already exists in my dictionary. If so, I will increment the count of how many times that URL has been encountered by 1; otherwise, I'll introduce a new dictionary entry for that URL and initialize it to the value of 1. I do that for every line in the log, sort the results in reverse order, numerically, and print them out:

```
URLCounts = {}
with open(logPath, "r") as f:
    for line in (l.rstrip() for l in f):
        match= format_pat.match(line)
        if match:
            access = match.groupdict()
            request = access['request']
            (action, URL, protocol) = request.split()
            if URLCounts.has_key(URL):
                URLCounts[URL] = URLCounts[URL] + 1
            else:
                URLCounts[URL] = 1
results = sorted(URLCounts, key=lambda i: int(URLCounts[i]), reverse=True)

for result in results[:20]:
    print result + ": " + str(URLCounts[result])
```

So, let's go ahead and run that:

```
IOErrorTraceback (most recent call last)
<ipython-input-3-281d53278f3c> in <module>()
      1 URLCounts = {}
      2
----> 3 with open(logPath, "r") as f:
      4     for line in (l.rstrip() for l in f):
      5         match= format_pat.match(line)

IOError: [Errno 2] No such file or directory: 'E:\\sundog-consult\\Udemy\\DataScience\\access_log.txt'
```

Oops! We end up with this big old error here. It's telling us that, we need more than 1 value to unpack. So apparently, we're getting some requests fields that don't contain an action, a URL, and a protocol that they contain something else.

Dealing with Real-World Data

Let's see what's going on there! So, if we print out all the requests that don't contain three items, we'll see what's actually showing up. So, what we're going to do here is a similar little snippet of code, but we're going to actually do that split on the request field, and print out cases where we don't get the expected three fields.

```
URLCounts = {}

with open(logPath, "r") as f:
    for line in (l.rstrip() for l in f):
        match= format_pat.match(line)
        if match:
            access = match.groupdict()
            request = access['request']
            fields = request.split()
            if (len(fields) != 3):
                print fields
```

Let's see what's actually in there:

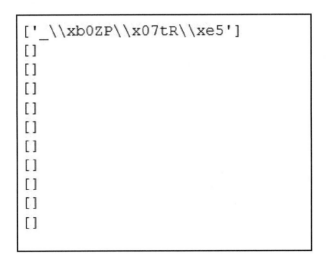

So, we have a bunch of empty fields. That's our first problem. But, then we have the first field that's full just garbage. Who knows where that came from, but it's clearly erroneous data. Okay, fine, let's modify our script.

Modification one - filtering the request field

We'll actually just throw out any lines that don't have the expected 3 fields in the request. That seems like a legitimate thing to do, because this does in fact have completely useless data inside of it, it's not like we're missing out on anything here by doing that. So, we'll modify our script to do that. We've introduced an `if (len(fields) == 3)` line before it actually tries to process it. We'll run that:

```
URLCounts = {}

with open(logPath, "r") as f:
    for line in (l.rstrip() for l in f):
        match= format_pat.match(line)
        if match:
            access = match.groupdict()
            request = access['request']
            fields = request.split()
            if (len(fields) == 3):
                URL = fields[1]
                if URLCounts.has_key(URL):
                    URLCounts[URL] = URLCounts[URL] + 1
                else:
                    URLCounts[URL] = 1

results = sorted(URLCounts, key=lambda i: int(URLCounts[i]), reverse=True)

for result in results[:20]:
    print result + ": " + str(URLCounts[result])
```

Dealing with Real-World Data

Hey, we got a result!

```
/xmlrpc.php: 68494
/wp-login.php: 1923
/: 440
/blog/: 138
/robots.txt: 123
/sitemap_index.xml: 118
/post-sitemap.xml: 118
/category-sitemap.xml: 117
/page-sitemap.xml: 117
/orlando-headlines/: 95
/san-jose-headlines/: 85
http://51.254.206.142/httptest.php: 81
/comics-2/: 76
/travel/: 74
/entertainment/: 72
/world/: 70
/business/: 70
/weather/: 70
/national/: 70
/national-headlines/: 70
```

But this doesn't really look like the top pages on my website. Remember, this is a news site. So, we're getting a bunch of PHP file hits, that's Perl scripts. What's going on there? Our top result is this `xmlrpc.php` script, and then `WP_login.php`, followed by the homepage. So, not very useful. Then there is `robots.txt`, then a bunch of XML files.

You know when I looked into this later on, it turned out that my site was actually under a malicious attack; someone was trying to break into it. This `xmlrpc.php` script was the way they were trying to guess at my passwords, and they were trying to log in using the login script. Fortunately, I shut them down before they could actually get through to this website.

This was an example of malicious data being introduced into my data stream that I have to filter out. So, by looking at that, we can see that not only was that malicious attack looking at PHP files, but it was also trying to execute stuff. It wasn't just doing a get request, it was doing a post request on the script to actually try to execute code on my website.

Modification two - filtering post requests

Now, I know that the data that I care about, you know in the spirit of the thing I'm trying to figure out is, people getting web pages from my website. So, a legitimate thing for me to do is to filter out anything that's not a get request, out of these logs. So, let's do that next. We're going to check again if we have three fields in our request field, and then we're also going to check if the action is get. If it's not, we're just going to ignore that line entirely:

```
URLCounts = {}

with open(logPath, "r") as f:
    for line in (l.rstrip() for l in f):
        match= format_pat.match(line)
        if match:
            access = match.groupdict()
            request = access['request']
            fields = request.split()
            if (len(fields) == 3):
                (action, URL, protocol) = fields
                if (action == 'GET'):
                    if URLCounts.has_key(URL):
                        URLCounts[URL] = URLCounts[URL] + 1
                    else:
                        URLCounts[URL] = 1

results = sorted(URLCounts, key=lambda i: int(URLCounts[i]), reverse=True)

for result in results[:20]:
    print result + ": " + str(URLCounts[result])
```

We should be getting closer to what we want now, the following is the output of the preceding code:

```
/: 434
/blog/: 138
/robots.txt: 123
/sitemap_index.xml: 118
/post-sitemap.xml: 118
/category-sitemap.xml: 117
/page-sitemap.xml: 117
/orlando-headlines/: 95
/san-jose-headlines/: 85
http://51.254.206.142/httptest.php: 81
/comics-2/: 76
/travel/: 74
/entertainment/: 72
/world/: 70
/business/: 70
/weather/: 70
/national/: 70
/national-headlines/: 70
/defense-sticking-head-sand/: 69
/about/: 69
```

Yeah, this is starting to look more reasonable. But, it still doesn't really pass a sanity check. This is a news website; people go to it to read news. Are they really reading my little blog on it that just has a couple of articles? I don't think so! That seems a little bit fishy. So, let's dive in a little bit, and see who's actually looking at those blog pages. If you were to actually go into that file and examine it by hand, you would see that a lot of these blog requests don't actually have any user agent on them. They just have a user agent of –, which is highly unusual:

```
54.165.199.171 - - [05/Dec/2015:09:32:05 +0000] "GET /blog/ HTTP/1.0" 200 31670 "-" "-"
```

If a real human being with a real browser was trying to get this page, it would say something like Mozilla, or Internet Explorer, or Chrome or something like that. So, it seems that these requests are coming from some sort of a scraper. Again, potentially malicious traffic that's not identifying who it is.

Modification three - checking the user agents

Maybe, we should be looking at the UserAgents too, to see if these are actual humans making requests, or not. Let's go ahead and print out all the different UserAgents that we're encountering. So, in the same spirit of the code that actually summed up the different URLs we were seeing, we can look at all the different UserAgents that we were seeing, and sort them by the most popular `user_agent` strings in this log:

```
UserAgents = {}

with open(logPath, "r") as f:
    for line in (l.rstrip() for l in f):
        match= format_pat.match(line)
        if match:
            access = match.groupdict()
            agent = access['user_agent']
            if UserAgents.has_key(agent):
                UserAgents[agent] = UserAgents[agent] + 1
            else:
                UserAgents[agent] = 1

results = sorted(UserAgents, key=lambda i: int(UserAgents[i]), reverse=True)

for result in results:
    print result + ": " + str(UserAgents[result])
```

We get the following result:

```
Mozilla/4.0 (compatible: MSIE 7.0; Windows NT 6.0): 68484
-: 4035
Mozilla/4.0 (compatible; MSIE 6.0; Windows NT 5.0): 1724
W3 Total Cache/0.9.4.1: 468
Mozilla/5.0 (compatible; Baiduspider/2.0; +http://www.baidu.com/search/spider.html): 278
Mozilla/5.0 (compatible; Googlebot/2.1; +http://www.google.com/bot.html): 248
Mozilla/5.0 (Windows NT 10.0; WOW64) AppleWebKit/537.36 (KHTML, like Gecko) Chrome/46.0.249
0.86 Safari/537.36: 158
Mozilla/5.0 (Windows NT 6.1; WOW64; rv:40.0) Gecko/20100101 Firefox/40.0: 144
Mozilla/5.0 (iPad; CPU OS 8_4 like Mac OS X) AppleWebKit/600.1.4 (KHTML, like Gecko) Versio
n/8.0 Mobile/12H143 Safari/600.1.4: 120
Mozilla/5.0 (Linux; Android 5.1.1; SM-G900T Build/LMY47X) AppleWebKit/537.36 (KHTML, like G
ecko) Chrome/46.0.2490.76 Mobile Safari/537.36: 47
Mozilla/5.0 (compatible; bingbot/2.0; +http://www.bing.com/bingbot.htm): 43
Mozilla/5.0 (compatible; MJ12bot/v1.4.5; http://www.majestic12.co.uk/bot.php?+): 41
Opera/9.80 (Windows NT 6.0) Presto/2.12.388 Version/12.14: 40
Mozilla/5.0 (compatible; YandexBot/3.0; +http://yandex.com/bots): 27
Ruby: 15
Mozilla/5.0 (Linux; Android 5.1.1; SM-G900T Build/LMY47X) AppleWebKit/537.36 (KHTML, like G
```

You can see most of it looks legitimate. So, if it's a scraper, and in this case it actually was a malicious attack but they were actually pretending to be a legitimate browser. But this dash `user_agent` shows up a lot too. So, I don't know what that is, but I know that it isn't an actual browser.

The other thing I'm seeing is a bunch of traffic from spiders, from web crawlers. So, there is Baidu which is a search engine in China, there is Googlebot just crawling the page. I think I saw Yandex in there too, a Russian search engine. So, our data is being polluted by a lot of crawlers that are just trying to mine our website for search engine purposes. Again, that traffic shouldn't count toward the intended purpose of my analysis, of seeing what pages these actual human beings are looking at on my website. These are all automated scripts.

Filtering the activity of spiders/robots

Alright, so this gets a little bit tricky. There's no real good way of identifying spiders or robots just based on the user string alone. But we can at least take a legitimate crack at it, and filter out anything that has the word "bot" in it, or anything from my caching plugin that might be requesting pages in advance as well. We'll also strip out our friend single dash. So, we will once again refine our script to, in addition to everything else, strip out any UserAgents that look fishy:

```
URLCounts = {}

with open(logPath, "r") as f:
    for line in (l.rstrip() for l in f):
        match= format_pat.match(line)
        if match:
            access = match.groupdict()
            agent = access['user_agent']
            if (not('bot' in agent or 'spider' in agent or
                    'Bot' in agent or 'Spider' in agent or
                    'W3 Total Cache' in agent or agent =='-')):
                request = access['request']
                fields = request.split()
                if (len(fields) == 3):
                    (action, URL, protocol) = fields
                    if (action == 'GET'):
                        if URLCounts.has_key(URL):
                            URLCounts[URL] = URLCounts[URL] + 1
                        else:
                            URLCounts[URL] = 1

results = sorted(URLCounts, key=lambda i: int(URLCounts[i]), reverse=True)

for result in results[:20]:
    print result + ": " + str(URLCounts[result])

URLCounts = {}

with open(logPath, "r") as f:
    for line in (l.rstrip() for l in f):
        match= format_pat.match(line)
        if match:
            access = match.groupdict()
            agent = access['user_agent']
            if (not('bot' in agent or 'spider' in agent or
                    'Bot' in agent or 'Spider' in agent or
                    'W3 Total Cache' in agent or agent =='-')):
                request = access['request']
```

Dealing with Real-World Data

```
            fields = request.split()
            if (len(fields) == 3):
                (action, URL, protocol) = fields
                if (URL.endswith("/")):
                    if (action == 'GET'):
                        if URLCounts.has_key(URL):
                            URLCounts[URL] = URLCounts[URL] + 1
                        else:
                            URLCounts[URL] = 1

results = sorted(URLCounts, key=lambda i: int(URLCounts[i]), reverse=True)

for result in results[:20]:
    print result + ": " + str(URLCounts[result])
```

What do we get?

```
/: 77
/orlando-headlines/: 36
/?page_id=34248: 28
/wp-content/cache/minify/000000/M9bPKixNLarUy00szs8D0Z15AA.js: 27
/wp-content/cache/minify/000000/lY7dDoIwDIVfiGOKxkfxfnbdKO4HuxICTy-it8Zw15PzfSftzPCckJem-x4qUWArqBP15myqZLEgyhdOaoxTo
GyGaiAL1OfUnIz0qDLOdSZGE-nOlpc3kopDzrSyavVVt_veb5qSDVhjsQ6dHh_B_eE_z2pYIGJ7iBWKeEio_eT9UQe4xHhDll27mGRryVu_pRc.js: 27
/wp-content/cache/minify/000000/M9AvyUjVzUstLy7PLErVz8lMKkosqtTPRtYvTi7KLCgpBgA.js: 27
/wp-content/cache/minify/000000/fY45DoAwDAQ_FMvkRQgFA5ZyWLajiN9zNHR0O83MRkyt-pIctqYFJPedKyYzfHg2PzOFiENAzaD07AxcpKmTo
1ORvDjZt8KEfhBUGjZYCf8FbOfvAlTXCw.css: 25
/?author=1: 21
/wp-content/cache/minify/000000/hcrRCYAwDAXAhXyEjiQ1YKAh4SVSx3cE7_uG7ASr4M9qg3kGWykladklK84LHtRj_My6YOPfqcz-AA.js: 20
/wp-content/uploads/2014/11/nhnl.png: 19
/wp-includes/js/wp-emoji-release.min.js?ver=4.3.1: 17
/wp-content/cache/minify/000000/BcGBCQAgCATAiUSaKYSERPk3avzuht4SkBJnt4tHJdqgnPBqKldesTcN1R8.js: 17
/wp-login.php: 16
/comics-2/: 12
/world/: 12
/favicon.ico: 10
/wp-content/uploads/2014/11/babyblues.jpg: 6
/wp-content/uploads/2014/11/garfield.jpg: 6
/wp-content/uploads/2014/11/violentcrime.jpg: 6
/robots.txt: 6
```

Alright, so here we go! This is starting to look more reasonable for the first two entries, the homepage is most popular, which would be expected. Orlando headlines is also popular, because I use this website more than anybody else, and I live in Orlando. But after that, we get a bunch of stuff that aren't webpages at all: a bunch of scripts, a bunch of CSS files. Those aren't web pages.

Modification four - applying website-specific filters

I can just apply some knowledge about my site, where I happen to know that all the legitimate pages on my site just end with a slash in their URL. So, let's go ahead and modify this again, to strip out anything that doesn't end with a slash:

```
URLCounts = {}

with open (logPath, "r") as f:
    for line in (l.rstrip() for l in f):
        match= format_pat.match(line)
        if match:
            access = match.groupdict()
            agent = access['user_agent']
            if (not('bot' in agent or 'spider' in agent or
                    'Bot' in agent or 'Spider' in agent or
                    'W3 Total Cache' in agent or agent =='-')):
                request = access['request']
                fields = request.split()
                if (len(fields) == 3):
                    (action, URL, protocol) = fields
                    if (URL.endswith("/")):
                        if (action == 'GET'):
                            if URLCounts.has_key(URL):
                                URLCounts[URL] = URLCounts[URL] + 1
                            else:
                                URLCounts[URL] = 1

results = sorted(URLCounts, key=lambda i: int(URLCounts[i]), reverse=True)

for result in results[:20]:
    print result + ": " + str(URLCounts[result])
```

Let's run that!

```
/: 77
/orlando-headlines/: 36
/world/: 12
/comics-2/: 12
/weather/: 4
/about/: 4
/australia/: 4
/national-headlines/: 3
/sample-page/feed/: 2
/feed/: 2
/technology/: 2
/science/: 2
/entertainment/: 1
/san-jose-headlines/: 1
/business/: 1
/travel/feed/: 1
```

Finally, we're getting some results that seem to make sense! So, it looks like, that the top page requested from actual human beings on my little No-Hate News site is the homepage, followed by `orlando-headlines`, followed by world news, followed by the comics, then the weather, and the about screen. So, this is starting to look more legitimate.

If you were to dig even deeper though, you'd see that there are still problems with this analysis. For example, those feed pages are still coming from robots just trying to get RSS data from my website. So, this is a great parable in how a seemingly simple analysis requires a huge amount of pre-processing and cleaning of the source data before you get results that make any sense.

Again, make sure the things you're doing to clean your data along the way are principled, and you're not just cherry-picking problems that don't match with your preconceived notions. So, always question your results, always look at your source data, and look for weird things that are in it.

Activity for web log data

Alright, if you want to mess with this some more you can solve that feed problem. Go ahead and strip out things that include feed because we know that's not a real web page, just to get some familiarity with the code. Or, go look at the log a little bit more closely, gain some understanding as to where those feed pages are actually coming from.

Maybe there's an even better and more robust way of identifying that traffic as a larger class. So, feel free to mess around with that. But I hope you learned your lesson: data cleaning - hugely important and it's going to take a lot of your time!

So, it's pretty surprising how hard it was to get some reasonable results on a simple question like "What are the top viewed pages on my website?" You can imagine if that much work had to go into cleaning the data for such a simple problem, think about all the nuanced ways that dirty data might actually impact the results of more complex problems, and complex algorithms.

It's very important to understand your source data, look at it, look at a representative sample of it, make sure you understand what's coming into your system. Always question your results and tie it back to the original source data to see where questionable results are coming from.

Normalizing numerical data

This is a very quick section: I just want to remind you about the importance of normalizing your data, making sure that your various input feature data is on the same scale, and is comparable. And, sometimes it matters, and sometimes it doesn't. But, you just have to be cognizant of when it does. Just keep that in the back of your head because sometimes it will affect the quality of your results if you don't.

So, sometimes models will be based on several different numerical attributes. If you remember multivariant models, we might have different attributes of a car that we're looking at, and they might not be directly comparable measurements. Or, for example, if we're looking at relationships between ages and incomes, ages might range from 0 to 100, but incomes in dollars might range from 0 to billions, and depending on the currency it could be an even larger range! Some models are okay with that.

If you're doing a regression, usually that's not a big deal. But, other models don't perform so well unless those values are scaled down first to a common scale. If you're not careful, you can end up with some attributes counting more than others. Maybe the income would end up counting much more than the age, if you were trying to treat those two values as comparable values in your model.

So this can introduce also a bias in the attributes, which can also be a problem. Maybe one set of your data is skewed, you know, sometimes you need to normalize things versus the actual range seen for that set of values and not just to a 0 to whatever the maximum is scale. There's no set rule as to when you should and shouldn't do this sort of normalization. All I can say is always read the documentation for whatever technique you're using.

So, for example, in scikit-learn their PCA implementation has a `whiten` option that will automatically normalize your data for you. You should probably use that. It also has some preprocessing modules available that will normalize and scale things for you automatically as well.

Be aware too of textual data that should actually be represented numerically, or ordinally. If you have `yes` or `no` data you might need to convert that to `1` or `0` and do that in a consistent matter. So again, just read the documentation. Most techniques do work fine with raw, un-normalized data, but before you start using a new technique for the first time, just read the documentation and understand whether or not the inputs should be scaled or normalized or whitened first. If so, scikit-learn will probably make it very easy for you to do so, you just have to remember to do it! Don't forget to rescale your results when you're done if you are scaling the input data.

If you want to be able to interpret the results you get, sometimes you need to scale them back up to their original range after you're done. If you are scaling things and maybe even biasing them towards a certain amount before you input them into a model, make sure that you unscale them and unbias them before you actually present those results to somebody. Or else they won't make any sense! And just a little reminder, a little bit of a parable if you will, always check to see if you should normalize or whiten your data before you pass it into a given model.

No exercise associated with this section; it's just something I want you to remember. I'm just trying to drive the point home. Some algorithms require whitening, or normalization, some don't. So, always read the documentation! If you do need to normalize the data going into an algorithm it will usually tell you so, and it will make it very easy to do so. Please just be aware of that!

Detecting outliers

A common problem with real-world data is outliers. You'll always have some strange users, or some strange agents that are polluting your data, that act abnormally and atypically from the typical user. They might be legitimate outliers; they might be caused by real people and not by some sort of malicious traffic, or fake data. So sometimes, it's appropriate to remove them, sometimes it isn't. Make sure you make that decision responsibly. So, let's dive into some examples of dealing with outliers.

For example, if I'm doing collaborative filtering, and I'm trying to make movie recommendations or something like that, you might have a few power users that have watched every movie ever made, and rated every movie ever made. They could end up having an inordinate influence on the recommendations for everybody else.

You don't really want a handful of people to have that much power in your system. So, that might be an example where it would be a legitimate thing to filter out an outlier, and identify them by how many ratings they've actually put into the system. Or, maybe an outlier would be someone who doesn't have enough ratings.

We might be looking at web log data, like we saw in our example earlier when we were doing data cleaning, outliers could be telling you that there's something very wrong with your data to begin with. It could be malicious traffic, it could be bots, or other agents that should be discarded that don't represent actual human beings that you're trying to model.

If someone really wanted the mean average income in the United States (and not the median), you shouldn't just throw out Donald Trump because you don't like him. You know the fact is, his billions of dollars are going to push that mean amount up, even if it doesn't budge the median. So, don't fudge your numbers by throwing out outliers. But throw out outliers if it's not consistent with what you're trying to model in the first place.

Now, how do we identify outliers? Well, remember our old friend standard deviation? We covered that very early in this book. It's a very useful tool for detecting outliers. You can, in a very principled matter, compute the standard deviation of a dataset that should have a more or less normal distribution. If you see a data point that's outside of one or two standard deviations, there you have an outlier.

Remember, we talked earlier too about the box and whisker diagrams too, and those also have a built-in way of detecting and visualizing outliers. Those diagrams define outliers as lying outside 1.5 the interquartile range.

Dealing with Real-World Data

What multiple do you choose? Well, you kind of have to use common sense, you know, there's no hard and fast rule as to what is an outlier. You have to look at your data and kind of eyeball it, look at the distribution, look at the histogram. See if there's actual things that stick out to you as obvious outliers, and understand what they are before you just throw them away.

Dealing with outliers

So, let's take some example code, and see how you might handle outliers in practice. Let's mess around with some outliers. It's a pretty simple section. A little bit of review actually. If you want to follow along, we're in `Outliers.ipynb`. So, go ahead and open that up if you'd like:

```
import numpy as np

incomes = np.random.normal(27000, 15000, 10000)
incomes = np.append(incomes, [1000000000])

import matplotlib.pyplot as plt
plt.hist(incomes, 50)
plt.show()
```

We did something very similar, very early in the book, where we created a fake histogram of income distribution in the United States. What we're going to do is start off with a normal distribution of incomes here that are have a mean of $27,000 per year, with a standard deviation of 15,000. I'm going to create 10,000 fake Americans that have an income in that distribution. This is totally made-up data, by the way, although it's not that far off from reality.

Then, I'm going to stick in an outlier - call it Donald Trump, who has a billion dollars. We're going to stick this guy in at the end of our dataset. So, we have a normally distributed dataset around $27,000, and then we're going to stick in Donald Trump at the end.

We'll go ahead and plot that as a histogram:

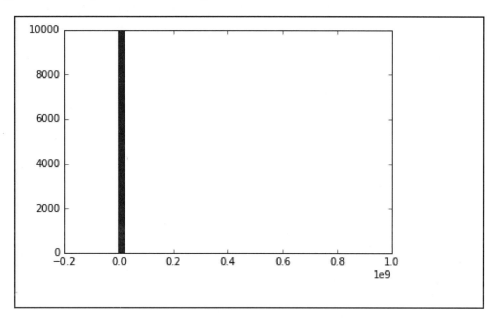

Wow! That's not very helpful! We have the entire normal distribution of everyone else in the country squeezed into one bucket of the histogram. On the other hand, we have Donald Trump out at the right side screwing up the whole thing at a billion dollars.

The other problem too is that if I'm trying to answer the question how much money does the typical American make. If I take the mean to try and figure that out, it's not going to be a very good, useful number:

```
incomes.mean ()
```

The output of the preceding code is as follows:

```
126892.66469341301
```

Donald Trump has pushed that number up all by himself to $126,000 and some odd of change, when I know that the real mean of my normally distributed data that excludes Donald Trump is only $27,000. So, the right thing to do there would be to use the median instead of the mean.

Dealing with Real-World Data

But, let's say we had to use the mean for some reason, and the right way to deal with this would be to exclude these outliers like Donald Trump. So, we need to figure out how do we identify these people. Well, you could just pick some arbitrary cutoff, and say, "I'm going to throw out all the billionaires", but that's not a very principled thing to do. Where did 1 billion come from?

It's just some accident of how we count numbers. So, a better thing to do would be to actually measure the standard deviation of your dataset, and identify outliers as being some multiple of a standard deviation away from the mean.

So, following is a little function that I wrote that does just that. It's called `reject_outliers()`:

```
def reject_outliers(data):
    u = np.median(data)
    s = np.std(data)
    filtered = [e for e in data if (u - 2 * s < e < u + 2 * s)]
    return filtered

filtered = reject_outliers(incomes)

plt.hist(filtered, 50)
plt.show()
```

It takes in a list of data and finds the median. It also finds the standard deviation of that dataset. So, I filter that out, so I only preserve data points that are within two standard deviations of the median for my data. So, I can use this handy dandy `reject_outliers()` function on my income data, to actually strip out weird outliers automatically:

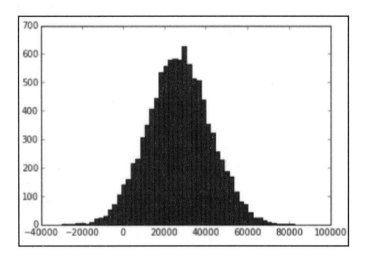

Sure enough, it works! I get a much prettier graph now that excludes Donald Trump and focuses in on the more typical dataset here in the center. So, pretty cool stuff!

So, that's one example of identifying outliers, and automatically removing them, or dealing with them however you see fit. Remember, always do this in a principled manner. Don't just throw out outliers because they're inconvenient. Understand where they're coming from, and how they actually affect the thing you're trying to measure in spirit.

By the way, our mean is also much more meaningful now; much closer to 27,000 that it should be, now that we've gotten rid of that outlier.

Activity for outliers

So, if you want to play around with this, you know just fiddle around with it like I normally ask you to do. Try different multiples of the standard deviation, try adding in more outliers, try adding in outliers that aren't quite as outlier-ish as Donald Trump. You know, just fabricate some extra fake data there and play around with it, see if you can identify those people successfully.

So there you have it! Outliers; pretty simple concept. So, that's an example of identifying outliers by looking at standard deviations, and just looking at the number of standard deviations from the mean or median that you care about. Median is probably a better choice actually, given that the outliers might be skewing the mean in and of themselves, right? So, by using the standard deviation, that's a good way of identifying outliers in a more principled manner than just picking some arbitrary cutoff. Again, you need to decide what the right thing to do is with those outliers. What are you actually trying to measure? Is it appropriate to actually discard them or not? So, keep that in your head!

Summary

In this chapter, we talked about the importance of striking a balance between bias and variance and minimizing error. Next, we saw the concept of k-fold cross-validation and how to implement it in Python to prevent overfitting. We learned the importance of cleaning data and normalizing it before processing it. We then saw an example to determine the popular pages of a website. In `Chapter 9`, *Apache Spark - Machine Learning on Big Data* we'll learn machine learning on big data using Apache Spark.

9
Apache Spark - Machine Learning on Big Data

So far in this book we've talked about a lot of general data mining and machine learning techniques that you can use in your data science career, but they've all been running on your desktop. As such, you can only run as much data as a single machine can process using technologies such as Python and scikit-learn.

Now, everyone talks about big data, and odds are you might be working for a company that does in fact have big data to process. Big data meaning that you can't actually control it all, you can't actually wrangle it all on just one system. You need to actually compute it using the resources of an entire cloud, a cluster of computing resources. And that's where Apache Spark comes in. Apache Spark is a very powerful tool for managing big data, and doing machine learning on large Datasets. By the end of the chapter, you will have an in-depth knowledge of the following topics:

- Installing and working with Spark
- **Resilient Distributed Datasets (RDDs)**
- The **MLlib (Machine Learning Library)**
- Decision Trees in Spark
- K-Means Clustering in Spark

Installing Spark

In this section, I'm going to get you set up using Apache Spark, and show you some examples of actually using Apache Spark to solve some of the same problems that we solved using a single computer in the past in this book. The first thing we need to do is get Spark set up on your computer. So, we're going to walk you through how to do that in the next couple of sections. It's pretty straightforward stuff, but there are a few gotchas. So, don't just skip these sections; there are a few things you need to pay special attention to get Spark running successfully, especially on a Windows system. Let's get Apache Spark set up on your system, so you can actually dive in and start playing around with it.

We're going to be running this just on your own desktop for now. But, the same programs that we're going to write in this chapter could be run on an actual Hadoop cluster. So, you can take these scripts that we're writing and running locally on your desktop in Spark standalone mode, and actually run them from the master node of an actual Hadoop cluster, then let it scale up to the entire power of a Hadoop cluster and process massive Datasets that way. Even though we're going to set things up to run locally on your own computer, keep in mind that these same concepts will scale up to running on a cluster as well.

Installing Spark on Windows

Getting Spark installed on Windows involves several steps that we'll walk you through here. I'm just going to assume that you're on Windows because most people use this book at home. We'll talk a little bit about dealing with other operating systems in a moment. If you're already familiar with installing stuff and dealing with environment variables on your computer, then you can just take the following little cheat sheet and go off and do it. If you're not so familiar with Windows internals, I will walk you through it one step at a time in the upcoming sections. Here are the quick steps for those Windows pros:

1. **Install a JDK**: You need to first install a JDK, that's a Java Development Kit. You can just go to Sun's website and download that and install it if you need to. We need the JDK because, even though we're going to be developing in Python during this course, that gets translated under the hood to Scala code, which is what Spark is developed in natively. And, Scala, in turn, runs on top of the Java interpreter. So, in order to run Python code, you need a Scala system, which will be installed by default as part of Spark. Also, we need Java, or more specifically Java's interpreter, to actually run that Scala code. It's like a technology layer cake.

2. **Install Python**: Obviously you're going to need Python, but if you've gotten to this point in the book, you should already have a Python environment set up, hopefully with Enthought Canopy. So, we can skip this step.

3. **Install a prebuilt version of Spark for Hadoop**: Fortunately, the Apache website makes available prebuilt versions of Spark that will just run out of the box that are precompiled for the latest Hadoop version. You don't have to build anything, you can just download that to your computer and stick it in the right place and be good to go for the most part.
4. **Create a conf/log4j.properties file**: We have a few configuration things to take care of. One thing we want to do is adjust our warning level so we don't get a bunch of warning spam when we run our jobs. We'll walk through how to do that. Basically, you need to rename one of the properties files, and then adjust the error setting within it.
5. **Add a SPARK_HOME environment variable**: Next, we need to set up some environment variables to make sure that you can actually run Spark from any path that you might have. We're going to add a SPARK_HOME environment variable pointing to where you installed Spark, and then we will add `%SPARK_HOME%\bin` to your system path, so that when you run Spark Submit, or PySpark or whatever Spark command you need, Windows will know where to find it.
6. **Set a HADOOP_HOME variable**: On Windows there's one more thing we need to do, we need to set a `HADOOP_HOME` variable as well because it's going to expect to find one little bit of Hadoop, even if you're not using Hadoop on your standalone system.
7. **Install winutils.exe**: Finally, we need to install a file called `winutils.exe`. There's a link to `winutils.exe` within the resources for this book, so you can get that there.

If you want to walk through the steps in more detail, you can refer to the upcoming sections.

Installing Spark on other operating systems

A quick note on installing Spark on other operating systems: the same steps will basically apply on them too. The main difference is going to be in how you set environment variables on your system, in such a way that they will automatically be applied whenever you log in. That's going to vary from OS to OS. macOS does it differently from various flavors of Linux, so you're going to have to be at least a little bit familiar with using a Unix terminal command prompt, and how to manipulate your environment to do that. But most macOS or Linux users who are doing development already have those fundamentals under their belt. And of course, you're not going to need `winutils.exe` if you're not on Windows. So, those are the main differences for installing on different OSes.

Installing the Java Development Kit

For installing the Java Development Kit, go back to the browser, open a new tab, and just search for `jdk` (short for Java Development Kit). This will bring you to the Oracle site, from where you can download Java:

On the Oracle website, click on JDK DOWNLOAD. Now, click on **Accept License Agreement** and then you can select the download option for your operating system:

Java Platform, Standard Edition	
Java SE 8u131 Java SE 8u131 includes important security fixes and bug fixes. Oracle strongly recommends that all Java SE 8 users upgrade to this release. Learn more ▸ **Important planned change for MD5-signed JARs** Starting with the April Critical Patch Update releases, planned for April 18 2017, all JRE versions will treat JARs signed with MD5 as unsigned. Learn more and view testing instructions. For more information on cryptographic algorithm support, please check the JRE and JDK Crypto Roadmap.	
<ul><li>Installation Instructions</li><li>Release Notes</li><li>Oracle License</li><li>Java SE Products</li><li>Third Party Licenses</li><li>Certified System Configurations</li><li>Readme Files<ul><li>JDK ReadMe</li><li>JRE ReadMe</li></ul></li></ul>	**JDK** DOWNLOAD ⬇ **Server JRE** DOWNLOAD ⬇ **JRE** DOWNLOAD ⬇

For me, that's going to be Windows 64-bit, and a wait for 198 MB of goodness to download:

Java SE Development Kit 8u131

You must accept the Oracle Binary Code License Agreement for Java SE to download this software.
Thank you for accepting the Oracle Binary Code License Agreement for Java SE; you may now download this software.

Product / File Description	File Size	Download
Linux ARM 32 Hard Float ABI	77.87 MB	jdk-8u131-linux-arm32-vfp-hflt.tar.gz
Linux ARM 64 Hard Float ABI	74.81 MB	jdk-8u131-linux-arm64-vfp-hflt.tar.gz
Linux x86	164.66 MB	jdk-8u131-linux-i586.rpm
Linux x86	179.39 MB	jdk-8u131-linux-i586.tar.gz
Linux x64	162.11 MB	jdk-8u131-linux-x64.rpm
Linux x64	176.95 MB	jdk-8u131-linux-x64.tar.gz
Mac OS X	226.57 MB	jdk-8u131-macosx-x64.dmg
Solaris SPARC 64-bit	139.79 MB	jdk-8u131-solaris-sparcv9.tar.Z
Solaris SPARC 64-bit	99.13 MB	jdk-8u131-solaris-sparcv9.tar.gz
Solaris x64	140.51 MB	jdk-8u131-solaris-x64.tar.Z
Solaris x64	96.96 MB	jdk-8u131-solaris-x64.tar.gz
Windows x86	191.22 MB	jdk-8u131-windows-i586.exe
Windows x64	198.03 MB	jdk-8u131-windows-x64.exe

Once the download is finished, locate the installer and start it running. Note that we can't just accept the default settings in the installer on Windows here. So, this is a Windows-specific workaround, but as of the writing of this book, the current version of Spark is 2.1.1 and it turns out there's an issue with Spark 2.1.1 with Java on Windows. The issue is that if you've installed Java to a path that has a space in it, it doesn't work, so we need to make sure that Java is installed to a path that does not have a space in it. This means that you can't skip this step even if you have Java installed already, so let me show you how to do that. On the installer, click on **Next**, and you will see, as in the following screen, that it wants to install by default to the `C:\Program Files\Java\jdk` path, whatever the version is:

Chapter 9

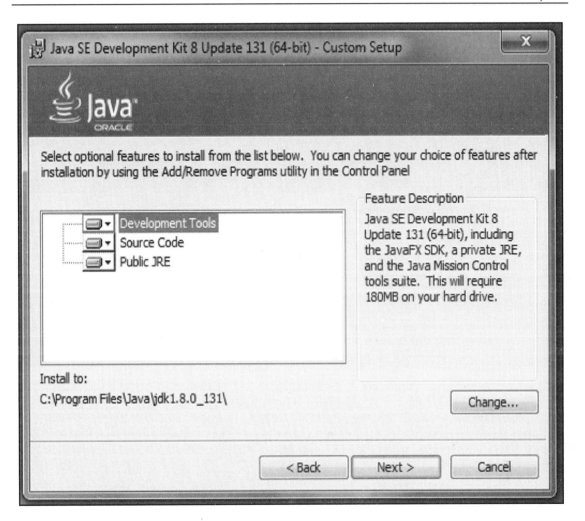

Apache Spark - Machine Learning on Big Data

The space in the `Program Files` path is going to cause trouble, so let's click on the **Change...** button and install to `c:\jdk`, a nice simple path, easy to remember, and with no spaces in it:

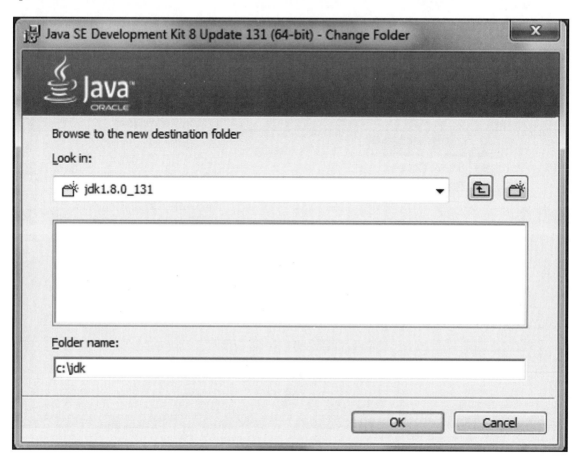

Now, it also wants to install the Java Runtime environment, so just to be safe, I'm also going to install that to a path with no spaces.

Chapter 9

At the second step of the JDK installation, we should have this showing on our screen:

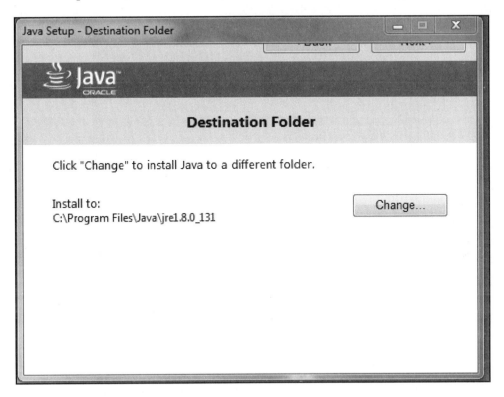

I will change that destination folder as well, and we will make a new folder called `C:\jre` for that:

Alright, successfully installed. Woohoo!

Now, you'll need to remember the path that we installed the JDK into, which in our case was `C:\jdk`. We still have a few more steps to go here. Next, we need to install Spark itself.

Installing Spark

Let's get back to a new browser tab here, head to `spark.apache.org`, and click on the **Download Spark** button:

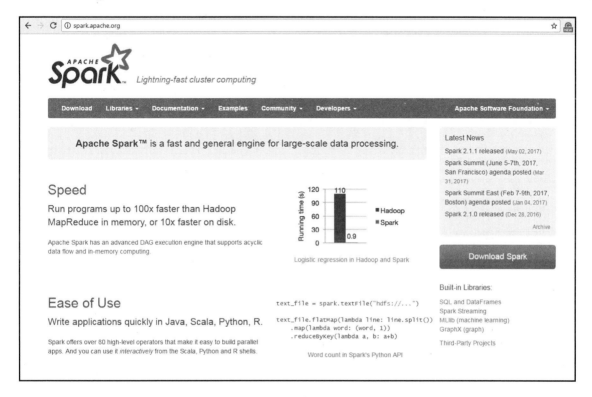

Now, we have used Spark 2.1.1 in this book, but anything beyond 2.0 should work just fine.

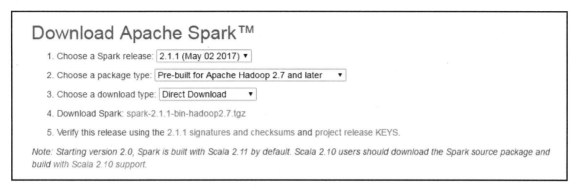

Make sure you get a prebuilt version, and select the **Direct Download** option so all these defaults are perfectly fine. Go ahead and click on the link next to instruction number **4** to download that package.

Now, it downloads a **TGZ** (**Tar in GZip**) file, which you might not be familiar with. Windows is kind of an afterthought with Spark quite honestly because on Windows, you're not going to have a built-in utility for actually decompressing TGZ files. This means that you might need to install one, if you don't have one already. The one I use is called WinRAR, and you can pick that up from `www.rarlab.com`. Go to the **Downloads** page if you need it, and download the installer for WinRAR 32-bit or 64-bit, depending on your operating system. Install WinRAR as normal, and that will allow you to actually decompress TGZ files on Windows:

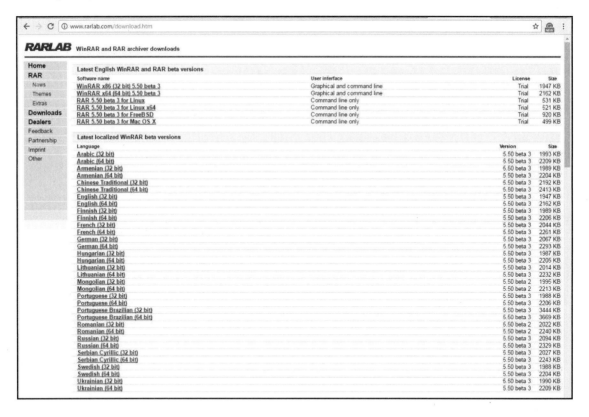

Chapter 9

So, let's go ahead and decompress the TGZ files. I'm going to open up my `Downloads` folder to find the Spark archive that we downloaded, and let's go ahead and right-click on that archive and extract it to a folder of my choosing - I'm just going to put it in my `Downloads` folder for now. Again, WinRAR is doing this for me at this point:

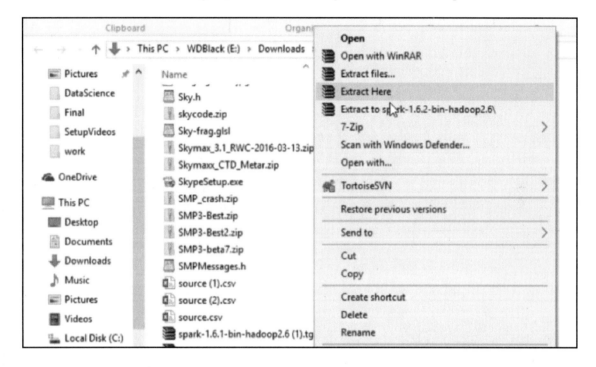

So, I should now have a folder in my `Downloads` folder associated with that package. Let's open that up and there is Spark itself. You should see something like the folder content shown below. So, you need to install that in some place that you can remember:

Name	Date modified	Type	Size
bin	6/1/2017 6:42 PM	File folder	
conf	6/1/2017 6:42 PM	File folder	
data	6/1/2017 6:42 PM	File folder	
examples	6/1/2017 6:42 PM	File folder	
jars	6/1/2017 6:42 PM	File folder	
licenses	6/1/2017 6:42 PM	File folder	
python	6/1/2017 6:42 PM	File folder	
R	6/1/2017 6:42 PM	File folder	
sbin	6/1/2017 6:42 PM	File folder	
yarn	6/1/2017 6:42 PM	File folder	
LICENSE	4/26/2017 5:40 AM	File	18 KB
NOTICE	4/26/2017 5:40 AM	File	25 KB
RELEASE	4/26/2017 5:40 AM	File	1 KB

You don't want to leave it in your `Downloads` folder obviously, so let's go ahead and open up a new file explorer window here. I go to my `C` drive and create a new folder, and let's just call it `spark`. So, my Spark installation is going to live in `C:\spark`. Again, nice and easy to remember. Open that folder. Now, I go back to my downloaded `spark` folder and use *Ctrl + A* to select everything in the Spark distribution, *Ctrl + C* to copy it, and then go back to `C:\spark`, where I want to put it, and *Ctrl + V* to paste it in:

Chapter 9

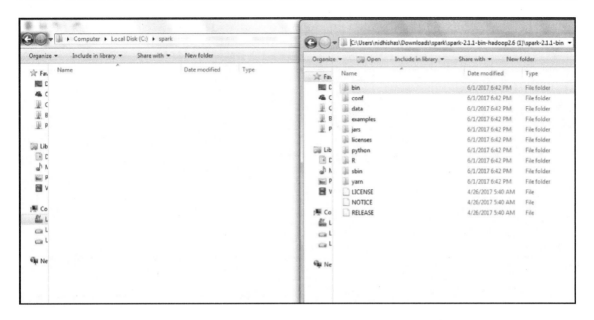

Remembering to paste the contents of the spark folder, not the spark folder itself is very important. So, what I should have now is my C drive with a spark folder that contains all of the files and folders from the Spark distribution.

Well, there are still a few things we need to configure. So, while we're in C:\spark let's open up the conf folder, and in order to make sure that we don't get spammed to death by log messages, we're going to change the logging level setting here. So to do that, right-click on the log4j.properties.template file and select **Rename**:

docker.properties.template	4/26/2017 5:40 AM	TEMPLATE File	1 KB
fairscheduler.xml.template	4/26/2017 5:40 AM	TEMPLATE File	2 KB
log4j.properties.template	4/26/2017 5:40 AM	TEMPLATE File	2 KB
metrics.properties.template	4/26/2017 5:40 AM	TEMPLATE File	8 KB
slaves.template	4/26/2017 5:40 AM	TEMPLATE File	1 KB
spark-defaults.conf.template	4/26/2017 5:40 AM	TEMPLATE File	2 KB
spark-env.sh.template	4/26/2017 5:40 AM	TEMPLATE File	4 KB

Delete the `.template` part of the filename to make it an actual `log4j.properties` file. Spark will use this to configure its logging:

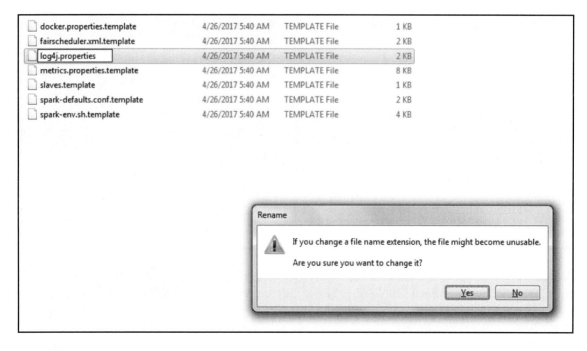

Now, open this file in a text editor of some sort. On Windows, you might need to right-click there and select **Open with** and then **WordPad**. In the file, locate `log4j.rootCategory=INFO`:

```
18  # Set everything to be logged to the console
19  log4j.rootCategory=INFO, console
20  log4j.appender.console=org.apache.log4j.ConsoleAppender
21  log4j.appender.console.target=System.err
22  log4j.appender.console.layout=org.apache.log4j.PatternLayout
23  log4j.appender.console.layout.ConversionPattern=%d{yy/MM/dd HH:mm:ss} %p %c{1}: %m%n
```

Let's change this to `log4j.rootCategory=ERROR` and this will just remove the clutter of all the log spam that gets printed out when we run stuff. Save the file, and exit your editor.

So far, we installed Python, Java, and Spark. Now the next thing we need to do is to install something that will trick your PC into thinking that Hadoop exists, and again this step is only necessary on Windows. So, you can skip this step if you're on Mac or Linux.

I have a little file available that will do the trick. Let's go to `http://media.sundog-soft.com/winutils.exe`. Downloading `winutils.exe` will give you a copy of a little snippet of an executable, which can be used to trick Spark into thinking that you actually have Hadoop:

Now, since we're going to be running our scripts locally on our desktop, it's not a big deal, we don't need to have Hadoop installed for real. This just gets around another quirk of running Spark on Windows. So, now that we have that, let's find it in the `Downloads` folder, *Ctrl* + *C* to copy it, and let's go to our `C` drive and create a place for it to live:

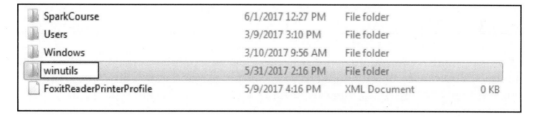

So, create a new folder again in the root `C` drive, and we will call it `winutils`:

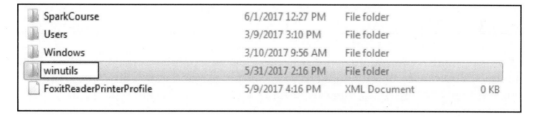

Now let's open this `winutils` folder and create a `bin` folder inside it:

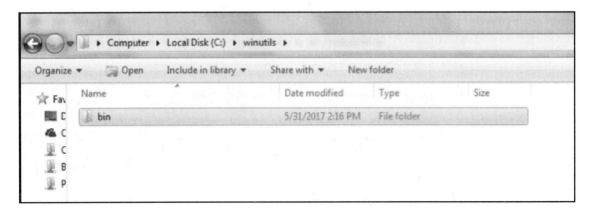

Now in this `bin` folder, I want you to paste the `winutils.exe` file we downloaded. So you should have `C:\winutils\bin` and then `winutils.exe`:

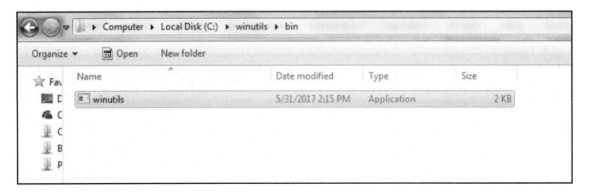

This next step is only required on some systems, but just to be safe, open Command Prompt on Windows. You can do that by going to your Start menu and going down to **Windows System**, and then clicking on **Command Prompt**. Here, I want you to type `cd c:\winutils\bin`, which is where we stuck our `winutils.exe` file. Now if you type `dir`, you should see that file there. Now type `winutils.exe chmod 777 \tmp\hive`. This just makes sure that all the file permissions you need to actually run Spark successfully are in place without any errors. You can close Command Prompt now that you're done with that step. Wow, we're almost done, believe it or not.

Now we need to set some environment variables for things to work. I'll show you how to do that on Windows. On Windows 10, you'll need to open up the Start menu and go to **Windows System | Control Panel** to open up **Control Panel**:

In **Control Panel**, click on **System and Security**:

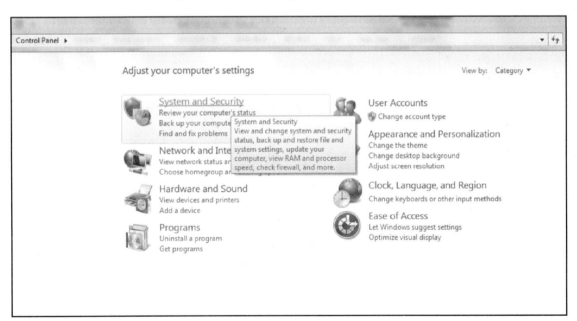

Then, click on **System**:

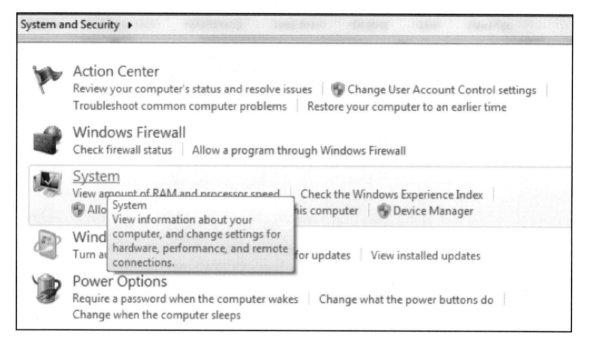

Then click on **Advanced system settings** from the list on the left-hand side:

From here, click on **Environment Variables...**:

We will get these options:

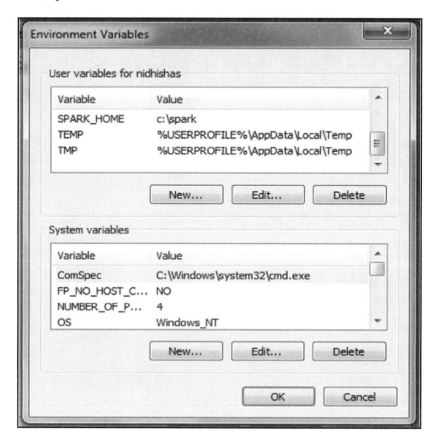

Now, this is a very Windows-specific way of setting environment variables. On other operating systems, you'll use different processes, so you'll have to look at how to install Spark on them. Here, we're going to set up some new user variables. Click on the first **New...** button for a new user variable and call it `SPARK_HOME`, as shown below, all uppercase. This is going to point to where we installed Spark, which for us is `c:\spark`, so type that in as the **Variable value** and click on **OK**:

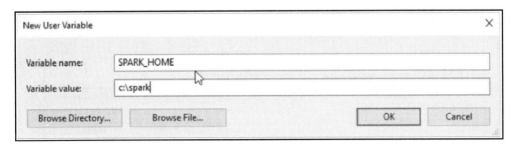

We also need to set up `JAVA_HOME`, so click on **New...** again and type in `JAVA_HOME` as **Variable name**. We need to point that to where we installed Java, which for us is `c:\jdk`:

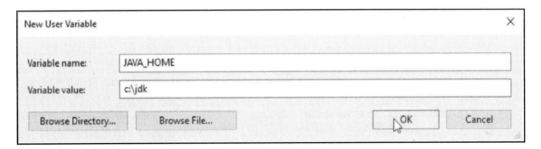

We also need to set up `HADOOP_HOME`, and that's where we installed the `winutils` package, so we'll point that to `c:\winutils`:

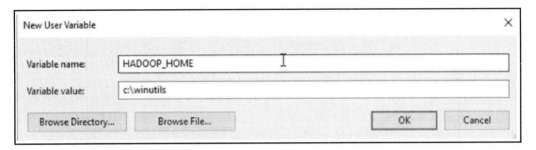

So far, so good. The last thing we need to do is to modify our path. You should have a **PATH** environment variable here:

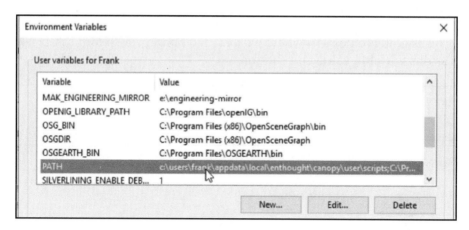

Click on the **PATH** environment variable, then on **Edit...**, and add a new path. This is going to be `%SPARK_HOME%\bin`, and I'm going to add another one, `%JAVA_HOME%\bin`:

Basically, this makes all the binary executables of Spark available to Windows, wherever you're running it from. Click on **OK** on this menu and on the previous two menus. We have finally everything set up.

Spark introduction

Let's get started with a high-level overview of Apache Spark and see what it's all about, what it's good for, and how it works.

What is Spark? Well, if you go to the Spark website, they give you a very high-level, hand-wavy answer, "A fast and general engine for large-scale data processing." It slices, it dices, it does your laundry. Well, not really. But it is a framework for writing jobs or scripts that can process very large amounts of data, and it manages distributing that processing across a cluster of computing for you. Basically, Spark works by letting you load your data into these large objects called Resilient Distributed Data stores, RDDs. It can automatically perform operations that transform and create actions based on those RDDs, which you can think of as large data frames.

The beauty of it is that Spark will automatically and optimally spread that processing out amongst an entire cluster of computers, if you have one available. You are no longer restricted to what you can do on a single machine or a single machine's memory. You can actually spread that out to all the processing capabilities and memory that's available to a cluster of machines, and, in this day and age, computing is pretty cheap. You can actually rent time on a cluster through things like Amazon's Elastic MapReduce service, and just rent some time on a whole cluster of computers for just a few dollars, and run your job that you couldn't run on your own desktop.

It's scalable

How is Spark scalable? Well, let's get a little bit more specific here in how it all works.

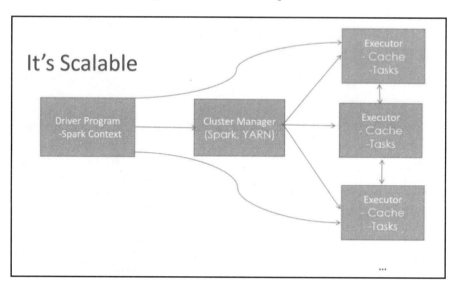

The way it works is, you write a driver program, which is just a little script that looks just like any other Python script really, and it uses the Spark library to actually write your script with. Within that library, you define what's called a Spark Context, which is sort of the root object that you work within when you're developing in Spark.

From there, the Spark framework kind of takes over and distributes things for you. So if you're running in standalone mode on your own computer, like we're going to be doing in these upcoming sections, it all just stays there on your computer, obviously. However, if you are running on a cluster manager, Spark can figure that out and automatically take advantage of it. Spark actually has its own built-in cluster manager, you can actually use it on its own without even having Hadoop installed, but if you do have a Hadoop cluster available to you, it can use that as well.

Hadoop is more than MapReduce; there's actually a component of Hadoop called YARN that separates out the entire cluster management piece of Hadoop. Spark can interface with YARN to actually use that to optimally distribute the components of your processing amongst the resources available to that Hadoop cluster.

Within a cluster, you might have individual executor tasks that are running. These might be running on different computers, or they might be running on different cores of the same computer. They each have their own individual cache and their own individual tasks that they run. The driver program, the Spark Context and the cluster manager work together to coordinate all this effort and return the final result back to you.

The beauty of it is, all you have to do is write the initial little script, the driver program, which uses a Spark Context to describe at a high level the processing you want to do on this data. Spark, working together with the cluster manager that you're using, figures out how to spread that out and distribute it so you don't have to worry about all those details. Well, if it doesn't work, obviously, you might have to do some troubleshooting to figure out if you have enough resources available for the task at hand, but, in theory, it's all just magic.

It's fast

What's the big deal about Spark? I mean, there are similar technologies like MapReduce that have been around longer. Spark is fast though, and on the website they claim that Spark is "up to 100x faster than MapReduce when running a job in memory, or 10 times faster on disk." Of course, the key words here are "up to," your mileage may vary. I don't think I've ever seen anything, actually, run that much faster than MapReduce. Some well-crafted MapReduce code can actually still be pretty darn efficient. But I will say that Spark does make a lot of common operations easier. MapReduce forces you to really break things down into mappers and reducers, whereas Spark is a little bit higher level. You don't have to always put as much thought into doing the right thing with Spark.

Part of that leads to another reason why Spark is so fast. It has a DAG engine, a directed acyclic graph. Wow, that's another fancy word. What does it mean? The way Spark works is, you write a script that describes how to process your data, and you might have an RDD that's basically like a data frame. You might do some sort of transformation on it, or some sort of action on it. But nothing actually happens until you actually perform an action on that data. What happens at that point is, Spark will say "hmm, OK. So, this is the end result you want on this data. What are all the other things I had to do to get up this point, and what's the optimal way to lay out the strategy for getting to that point?" So, under the hood, it will figure out the best way to split up that processing, and distribute that information to get the end result that you're looking for. So, the key inside here, is that Spark waits until you tell it to actually produce a result, and only at that point does it actually go and figure out how to produce that result. So, it's kind of a cool concept there, and that's the key to a lot of its efficiency.

It's young

Spark is a very hot technology, and is relatively young, so it's still very much emerging and changing quickly, but a lot of big people are using it. Amazon, for example, has claimed they're using it, eBay, NASA's Jet Propulsional Laboratories, Groupon, TripAdvisor, Yahoo, and many, many others have too. I'm sure there's a lot of companies using it that don't confess up to it, but if you go to the Spark Apache Wiki page at http://spark.apache.org/powered-by.html.

There's actually a list you can look up of known big companies that are using Spark to solve real-world data problems. If you are worried that you're getting into the bleeding edge here, fear not, you're in very good company with some very big people that are using Spark in production for solving real problems. It is pretty stable stuff at this point.

It's not difficult

It's also not that hard. You have your choice of programming in Python, Java, or Scala, and they're all built around the same concept that I just described earlier, that is, the Resilient Distributed Dataset, RDD for short. We'll talk about that in a lot more detail in the coming sections of this chapter.

Components of Spark

Spark actually has many different components that it's built up of. So there is a Spark Core that lets you do pretty much anything you can dream up just using Spark Core functions alone, but there are these other things built on top of Spark that are also useful.

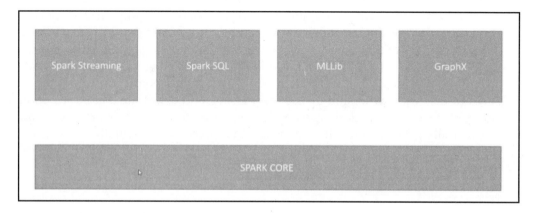

- **Spark Streaming**: Spark Streaming is a library that lets you actually process data in real time. Data can be flowing into a server continuously, say, from weblogs, and Spark Streaming can help you process that data in real time as you go, forever.
- **Spark SQL**: This lets you actually treat data as a SQL database, and actually issue SQL queries on it, which is kind of cool if you're familiar with SQL already.
- **MLlib**: This is what we're going to be focusing on in this section. It is actually a machine learning library that lets you perform common machine learning algorithms, with Spark underneath the hood to actually distribute that processing across a cluster. You can perform machine learning on much larger Datasets than you could have otherwise.
- **GraphX**: This is not for making pretty charts and graphs. It refers to graph in the network theory sense. Think about a social network; that's an example of a graph. GraphX just has a few functions that let you analyze the properties of a graph of information.

Python versus Scala for Spark

I do get some flack sometimes about using Python when I'm teaching people about Apache Spark, but there's a method to my madness. It is true that a lot of people use Scala when they're writing Spark code, because that's what Spark is developed in natively. So, you are incurring a little bit of overhead by forcing Spark to translate your Python code into Scala and then into Java interpreter commands at the end of the day.

However, Python's a lot easier, and you don't need to compile things. Managing dependencies is also a lot easier. You can really focus your time on the algorithms and what you're doing, and less on the minutiae of actually getting it built, and running, and compiling, and all that nonsense. Plus, obviously, this book has been focused on Python so far, and it makes sense to keep using what we've learned and stick with Python throughout these lectures. Here's a quick summary of the pros and cons of the two languages:

Python	Scala
• No need to compile, manage dependencies, etc. • Less coding overhead • You already know Python • Lets us focus on the concepts instead of a new language	• Scala is probably a more popular choice with Spark • Spark is built in Scala, so coding in Scala is "native" to Spark • New features, libraries tend to be Scala-first

However, I will say that if you were to do some Spark programming in the real world, there's a good chance people are using Scala. Don't worry about it too much, though, because in Spark the Python and Scala code ends up looking very similar because it's all around the same RDD concept. The syntax is very slightly different, but it's not that different. If you can figure out how to do Spark using Python, learning how to use it in Scala isn't that big of a leap, really. Here's a quick example of the same code in the two languages:

> **Python code to square numbers in a data set:**
>
> nums = sc.parallelize([1, 2, 3, 4])
> squared = nums.map(lambda x: x * x).collect()
>
> **Scala code to square numbers in a data set:**
>
> val nums = sc.parallelize(List(1, 2, 3, 4))
> val squared = nums.map(x => x * x).collect()

So, that's the basic concepts of Spark itself, why it's such a big deal, and how it's so powerful in letting you run machine learning algorithms on very large Datasets, or any algorithm really. Let's now talk in a little bit more detail about how it does that, and the core concept of the Resilient Distributed Dataset.

Spark and Resilient Distributed Datasets (RDD)

Let's get a little bit deeper into how Spark works. We're going to talk about Resilient Distributed Datasets, known as RDDs. It's sort of the core that you use when programming in Spark, and we'll have a few code snippets to try to make it real. We're going to give you a crash course in Apache Spark here. There's a lot more depth to it than what we're going to cover in the next few sections, but I'm just going to give you the basics you need to actually understand what's going on in these examples, and hopefully get you started and pointed in the right direction.

As mentioned, the most fundamental piece of Spark is called the Resilient Distributed Dataset, an RDD, and this is going to be the object that you use to actually load and transform and get the answers you want out of the data that you're trying to process. It's a very important thing to understand. The final letter in RDD stands for Dataset, and at the end of the day that's all it is; it's just a bunch of rows of information that can contain pretty much anything. But the key is the R and the first D.

- **Resilient**: It is resilient in that Spark makes sure that if you're running this on a cluster and one of those clusters goes down, it can automatically recover from that and retry. Now, that resilience only goes so far, mind you. If you don't have enough resources available to the job that you're trying to run, it will still fail, and you will have to add more resources to it. There's only so many things it can recover from; there is a limit to how many times it will retry a given task. But it does make its best effort to make sure that in the face of an unstable cluster or an unstable network it will still continue to try its best to run through to completion.
- **Distributed**: Obviously, it is distributed. The whole point of using Spark is that you can use it for big data problems where you can actually distribute the processing across the entire CPU and memory power of a cluster of computers. That can be distributed horizontally, so you can throw as many computers as you want to a given problem. The larger the problem, the more computers; there's really no upper bound to what you can do there.

The SparkContext object

You always start your Spark scripts by getting a SparkContext object, and this is the object that embodies the guts of Spark. It is what is going to give you your RDDs to process on, so it is what generates the objects that you use in your processing.

You know, you don't actually think about the SparkContext very much when you're actually writing Spark programs, but it is sort of the substrate that is running them for you under the hood. If you're running in the Spark shell interactively, it has an `sc` object already available for you that you can use to create RDDs. In a standalone script, however, you will have to create that SparkContext explicitly, and you'll have to pay attention to the parameters that you use because you can actually tell the Spark context how you want that to be distributed. Should I take advantage of every core that I have available to me? Should I be running on a cluster or just standalone on my local computer? So, that's where you set up the fundamental settings of how Spark will operate.

Creating RDDs

Let's look at some little code snippets of actually creating RDDs, and I think it will all start to make a little bit more sense.

Creating an RDD using a Python list

The following is a very simple example:

```
nums = parallelize([1, 2, 3, 4])
```

If I just want to make an RDD out of a plain old Python list, I can call the `parallelize()` function in Spark. That will convert a list of stuff, in this case, just the numbers, 1, 2, 3, 4, into an RDD object called `nums`.

That is the simplest case of creating an RDD, just from a hard-coded list of stuff. That list could come from anywhere; it doesn't have to be hard-coded either, but that kind of defeats the purpose of big data. I mean, if I have to load the entire Dataset into memory before I can create an RDD from it, what's the point?

Loading an RDD from a text file

I can also load an RDD from a text file, and that could be anywhere.

```
sc.textFile("file:///c:/users/frank/gobs-o-text.txt")
```

In this example, I have a giant text file that's the entire encyclopedia or something. I'm reading that from my local disk here, but I could also use s3n if I want to host this file on a distributed AmazonS3 bucket, or hdfs if I want to refer to data that's stored on a distributed HDFS cluster (that stands for Hadoop Distributed File System if you're not familiar with HDFS). When you're dealing with big data and working with a Hadoop cluster, usually that's where your data will live.

That line of code will actually convert every line of that text file into its own row in an RDD. So, you can think of the RDD as a database of rows, and, in this example, it will load up my text file into an RDD where every line, every row, contains one line of text. I can then do further processing in that RDD to parse or break out the delimiters in that data. But that's where I start from.

Remember when we talked about ETL and ELT earlier in the book? This is a good example of where you might actually be loading raw data into a system and doing the transform on the system itself that you used to query your data. You can take raw text files that haven't been processed at all and use the power of Spark to actually transform those into more structured data.

It can also talk to things like Hive, so if you have an existing Hive database set up at your company, you can create a Hive context that's based on your Spark context. How cool is that? Take a look at this example code:

```
hiveCtx = HiveContext(sc)  rows = hiveCtx.sql("SELECT name, age FROM users")
```

You can actually create an RDD, in this case called rows, that's generated by actually executing a SQL query on your Hive database.

More ways to create RDDs

There are more ways to create RDDs as well. You can create them from a JDBC connection. Basically any database that supports JDBC can also talk to Spark and have RDDs created from it. Cassandra, HBase, Elasticsearch, also files in JSON format, CSV format, sequence files object files, and a bunch of other compressed files like ORC can be used to create RDDs. I don't want to get into the details of all those, you can get a book and look those up if you need to, but the point is that it's very easy to create an RDD from data, wherever it might be, whether it's on a local filesystem or a distributed data store.

Again, RDD is just a way of loading and maintaining very large amounts of data and keeping track of it all at once. But, conceptually within your script, an RDD is just an object that contains a bunch of data. You don't have to think about the scale, because Spark does that for you.

RDD operations

Now, there are two different types of classes of things you can do on RDDs once you have them, you can do transformations, and you can do actions.

Transformations

Let's talk about transformations first. Transformations are exactly what they sound like. It's a way of taking an RDD and transforming every row in that RDD to a new value, based on a function you provide. Let's look at some of those functions:

- **map() and flatmap()**: `map` and `flatmap` are the functions you'll see the most often. Both of these will take any function that you can dream up, that will take, as input, a row of an RDD, and it will output a transformed row. For example, you might take raw input from a CSV file, and your `map` operation might take that input and break it up into individual fields based on the comma delimiter, and return back a Python list that has that data in a more structured format that you can perform further processing on. You can chain map operations together, so the output of one `map` might end up creating a new RDD that you then do another transformation on, and so on, and so forth. Again, the key is, Spark can distribute those transformations across the cluster, so it might take part of your RDD and transform it on one machine, and another part of your RDD and transform it on another.

    Like I said, `map` and `flatmap` are the most common transformations you'll see. The only difference is that `map` will only allow you to output one value for every row, whereas `flatmap` will let you actually output multiple new rows for a given row. So you can actually create a larger RDD or a smaller RDD than you started with using `flatmap`.

- **filter()**: `filter` can be used if what you want to do is just create a Boolean function that says "should this row be preserved or not? Yes or no."
- **distinct()**: `distinct` is a less commonly used transformation that will only return back distinct values within your RDD.
- **sample()**: This function lets you take a random sample from your RDD
- **union(), intersection(), subtract() and Cartesian()**: You can perform intersection operations like union, intersection, subtract, or even produce every cartesian combination that exists within an RDD.

Using map()

Here's a little example of how you might use the map function in your work:

```
rdd = sc.parallelize([1, 2, 3, 4])
rdd.map(lambda x: x*x)
```

Let's say I created an RDD just from the list 1, 2, 3, 4. I can then call `rdd.map()` with a lambda function of x that takes in each row, that is, each value of that RDD, calls it x, and then it applies the function x multiplied by x to square it. If I were to then collect the output of this RDD, it would be 1, 4, 9 and 16, because it would take each individual entry of that RDD and square it, and put that into a new RDD.

If you don't remember what lambda functions are, we did talk about it a little bit earlier in this book, but as a refresher, the lambda function is just a shorthand for defining a function in line. So `rdd.map(lambda x: x*x)` is exactly the same thing as a separate function `def squareIt(x): return x*x`, and saying `rdd.map(squareIt)`.

It's just a shorthand for very simple functions that you want to pass in as a transformation. It eliminates the need to actually declare this as a separate named function of its own. That's the whole idea of functional programming. So you can say you understand functional programming now, by the way! But really, it's just shorthand notation for defining a function inline as part of the parameters to a `map()` function, or any transformation for that matter.

Actions

You can also perform actions on an RDD, when you want to actually get a result. Here are some examples of what you can do:

- `collect()`: You can call collect() on an RDD, which will give you back a plain old Python object that you can then iterate through and print out the results, or save them to a file, or whatever you want to do.
- `count()`: You can also call `count()`, which will force it to actually go count how many entries are in the RDD at this point.
- `countByValue()`: This function will give you a breakdown of how many times each unique value within that RDD occurs.
- `take()`: You can also sample from the RDD using `take()`, which will take a random number of entries from the RDD.

- `top()`: `top()` will give you the first few entries in that RDD if you just want to get a little peek into what's in there for debugging purposes.
- `reduce()`: The more powerful action is `reduce()` which will actually let you combine values together for the same common key value. You can also use RDDs in the context of key-value data. The `reduce()` function lets you define a way of combining together all the values for a given key. It is very much similar in spirit to MapReduce. `reduce()` is basically the analogous operation to a `reducer()` in MapReduce, and `map()` is analogous to a `mapper()`. So, it's often very straightforward to actually take a MapReduce job and convert it to Spark by using these functions.

Remember, too, that nothing actually happens in Spark until you call an action. Once you call one of those action methods, that's when Spark goes out and does its magic with directed acyclic graphs, and actually computes the optimal way to get the answer you want. But remember, nothing really occurs until that action happens. So, that can sometimes trip you up when you're writing Spark scripts, because you might have a little print statement in there, and you might expect to get an answer, but it doesn't actually appear until the action is actually performed.

That is Spark 101 in a nutshell. Those are the basics you need for Spark programming. Basically, what is an RDD and what are the things you can do to an RDD. Once you get those concepts, then you can write some Spark code. Let's change tack now and talk about MLlib, and some specific features in Spark that let you do machine learning algorithms using Spark.

Introducing MLlib

Fortunately, you don't have to do things the hard way in Spark when you're doing machine learning. It has a built-in component called MLlib that lives on top of Spark Core, and this makes it very easy to perform complex machine learning algorithms using massive Datasets, and distributing that processing across an entire cluster of computers. So, very exciting stuff. Let's learn more about what it can do.

Some MLlib Capabilities

So, what are some of the things MLlib can do? Well, one is feature extraction.

One thing you can do at scale is term frequency and inverse document frequency stuff, and that's useful for creating, for example, search indexes. We will actually go through an example of that later in the chapter. The key, again, is that it can do this across a cluster using massive Datasets, so you could make your own search engine for the web with this, potentially. It also offers basic statistics functions, chi-squared tests, Pearson or Spearman correlation, and some simpler things like min, max, mean, and variance. Those aren't terribly exciting in and of themselves, but what is exciting is that you can actually compute the variance or the mean or whatever, or the correlation score, across a massive Dataset, and it would actually break that Dataset up into various chunks and run that across an entire cluster if necessary.

So, even if some of these operations aren't terribly interesting, what's interesting about it is the scale at which it can operate at. It can also support things like linear regression and logistic regression, so if you need to fit a function to a massive set of data and use that for predictions, you can do that too. It also supports Support Vector Machines. We're getting into some of the more fancy algorithms here, some of the more advanced stuff, and that too can scale up to massive Datasets using Spark's MLlib. There is a Naive Bayes classifier built into MLlib, so, remember that spam classifier that we built earlier in the book? You could actually do that for an entire e-mail system using Spark, and scale that up as far as you want to.

Decision trees, one of my favorite things in machine learning, are also supported by Spark, and we'll actually have an example of that later in this chapter. We'll also look at K-Means clustering, and you can do clustering using K-Means and massive Datasets with Spark and MLlib. Even principal component analysis and **SVD** (**Singular Value Decomposition**) can be done with Spark as well, and we'll have an example of that too. And, finally, there's a built-in recommendations algorithm called Alternating Least Squares that's built into MLlib. Personally, I've had kind of mixed results with it, you know, it's a little bit too much of a black box for my taste, but I am a recommender system snob, so take that with a grain of salt!

Special MLlib data types

Using MLlib is usually pretty straightforward, there are just some library functions you need to call. It does introduce a few new data types; however, that you need to know about, and one is the vector.

The vector data type

Remember when we were doing movie similarities and movie recommendations earlier in the book? An example of a vector might be a list of all the movies that a given user rated. There are two types of vector, sparse and dense. Let's look at an example of those. There are many, many movies in the world, and a dense vector would actually represent data for every single movie, whether or not a user actually watched it. So, for example, let's say I have a user who watched Toy Story, obviously I would store their rating for Toy Story, but if they didn't watch the movie Star Wars, I would actually store the fact that there is not a number for Star Wars. So, we end up taking up space for all these missing data points with a dense vector. A sparse vector only stores the data that exists, so it doesn't waste any memory space on missing data, OK. So, it's a more compact form of representing a vector internally, but obviously that introduces some complexity while processing. So, it's a good way to save memory if you know that your vectors are going to have a lot of missing data in them.

LabeledPoint data type

There's also a `LabeledPoint` data type that comes up, and that's just what it sounds like, a point that has some sort of label associated with it that conveys the meaning of this data in human readable terms.

Rating data type

Finally, there is a `Rating` data type that you'll encounter if you're using recommendations with MLlib. This data type can take in a rating that represents a 1-5 or 1-10, whatever star rating a person might have, and use that to inform product recommendations automatically.

So, I think you finally have everything you need to get started, let's dive in and actually look at some real MLlib code and run it, and then it will make a lot more sense.

Decision Trees in Spark with MLlib

Alright, let's actually build some decision trees using Spark and the MLlib library, this is very cool stuff. Wherever you put the course materials for this book, I want you to go to that folder now. Make sure you're completely closed out of Canopy, or whatever environment you're using for Python development, because I want to make sure you're starting it from this directory, OK? And find the `SparkDecisionTree` script, and double-click that to open up Canopy:

```python
from pyspark.mllib.regression import LabeledPoint
from pyspark.mllib.tree import DecisionTree
from pyspark import SparkConf, SparkContext
from numpy import array

# Boilerplate Spark stuff:
conf = SparkConf().setMaster("local").setAppName("SparkDecisionTree")
sc = SparkContext(conf = conf)

# Some functions that convert our CSV input data into numerical
# features for each job candidate
def binary(YN):
    if (YN == 'Y'):
        return 1
    else:
        return 0

def mapEducation(degree):
    if (degree == 'BS'):
        return 1
    elif (degree =='MS'):
        return 2
    elif (degree == 'PhD'):
        return 3
    else:
        return 0
```

Now, up until this point we've been using IPython notebooks for our code, but you can't really use those very well with Spark. With Spark scripts, you need to actually submit them to the Spark infrastructure and run them in a very special way, and we'll see how that works shortly.

Exploring decision trees code

So, we are just looking at a raw Python script file now, without any of the usual embellishment of the IPython notebook stuff. let's walk through what's going on in the script.

```python
from pyspark.mllib.regression import LabeledPoint
from pyspark.mllib.tree import DecisionTree
from pyspark import SparkConf, SparkContext
from numpy import array

# Boilerplate Spark stuff:
conf = SparkConf().setMaster("local").setAppName("SparkDecisionTree")
sc = SparkContext(conf = conf)

# Some functions that convert our CSV input data into numerical
# features for each job candidate
def binary(YN):
    if (YN == 'Y'):
        return 1
    else:
        return 0

def mapEducation(degree):
    if (degree == 'BS'):
        return 1
    elif (degree =='MS'):
        return 2
    elif (degree == 'PhD'):
        return 3
    else:
        return 0
```

We'll go through it slowly, because this is your first Spark script that you've seen in this book.

First, we're going to import, from `pyspark.mllib`, the bits that we need from the machine learning library for Spark.

```
from pyspark.mllib.regression import LabeledPoint
from pyspark.mllib.tree import DecisionTree
```

We need the `LabeledPoint` class, which is a data type required by the `DecisionTree` class, and the `DecisionTree` class itself, imported from `mllib.tree`.

Next, pretty much every Spark script you see is going to include this line, where we import `SparkConf` and `SparkContext`:

```
from pyspark import SparkConf, SparkContext
```

This is needed to create the `SparkContext` object that is kind of the root of everything you do in Spark.

And finally, we're going to import the array library from `numpy`:

```
from numpy import array
```

Yes, you can still use `NumPy`, and `scikit-learn`, and whatever you want within Spark scripts. You just have to make sure, first of all, that these libraries are installed on every machine that you intend to run it on.

If you're running on a cluster, you need to make sure that those Python libraries are already in place somehow, and you also need to understand that Spark will not make the scikit-learn methods, for example, magically scalable. You can still call these functions in the context of a given map function, or something like that, but it's only going to run on that one machine within that one process. Don't lean on that stuff too heavily, but, for simple things like managing arrays, it's totally an okay thing to do.

Creating the SparkContext

Now, we'll start by setting up our `SparkContext`, and giving it a `SparkConf`, a configuration.

```
conf = SparkConf().setMaster("local").setAppName("SparkDecisionTree")
```

This configuration object says, I'm going to set the master node to `"local"`, and this means that I'm just running on my own local desktop, I'm not actually running on a cluster at all, and I'm just going to run in one process. I'm also going to give it an app name of `"SparkDecisionTree"`, and you can call that whatever you want, Fred, Bob, Tim, whatever floats your boat. It's just what this job will appear as if you were to look at it in the Spark console later on.

Apache Spark - Machine Learning on Big Data

And then we will create our `SparkContext` object using that configuration:

```
sc = SparkContext(conf = conf)
```

That gives us an `sc` object we can use for creating RDDs.

Next, we have a bunch of functions:

```
# Some functions that convert our CSV input data into numerical
# features for each job candidate
def binary(YN):
    if (YN == 'Y'):
        return 1
    else:
        return 0

def mapEducation(degree):
    if (degree == 'BS'):
        return 1
    elif (degree =='MS'):
        return 2
    elif (degree == 'PhD'):
        return 3
    else:
        return 0

# Convert a list of raw fields from our CSV file to a
# LabeledPoint that MLLib can use. All data must be numerical...
def createLabeledPoints(fields):
    yearsExperience = int(fields[0])
    employed = binary(fields[1])
    previousEmployers = int(fields[2])
    educationLevel = mapEducation(fields[3])
    topTier = binary(fields[4])
    interned = binary(fields[5])
    hired = binary(fields[6])

    return LabeledPoint(hired, array([yearsExperience, employed,
        previousEmployers, educationLevel, topTier, interned]))
```

Let's just get down these functions for now, and we'll come back to them later.

Importing and cleaning our data

Let's go to the first bit of Python code that actually gets executed in this script.

```
43 rawData = sc.textFile("e:/sundog-consult/udemy/datascience/PastHires.csv")
44 header = rawData.first()
45 rawData = rawData.filter(lambda x:x != header)
46
47 # Split each line into a list based on the comma delimiters
48 csvData = rawData.map(lambda x: x.split(","))
49
50 # Convert these lists to LabeledPoints
51 trainingData = csvData.map(createLabeledPoints)
52
53 # Create a test candidate, with 10 years of experience, currently employed,
54 # 3 previous employers, a BS degree, but from a non-top-tier school where
55 # he or she did not do an internship. You could of course load up a whole
56 # huge RDD of test candidates from disk, too.
57 testCandidates = [ array([10, 1, 3, 1, 0, 0])]
58 testData = sc.parallelize(testCandidates)
```

The first thing we're going to do is load up this `PastHires.csv` file, and that's the same file we used in the decision tree exercise that we did earlier in this book.

Let's pause quickly to remind ourselves of the content of that file. If you remember right, we have a bunch of attributes of job candidates, and we have a field of whether or not we hired those people. What we're trying to do is build up a decision tree that will predict - would we hire or not hire a person given those attributes?

Now, let's take a quick peek at the `PastHires.csv`, which will be an Excel file.

	A	B	C	D	E	F	G
1	Years Exp	Employed	Previous	Level of E	Top-tier s	Interned	Hired
2	10	Y		4	BS	N	N	Y
3	0	N		0	BS	Y	Y	Y
4	7	N		6	BS	N	N	N
5	2	Y		1	MS	Y	N	Y
6	20	N		2	PhD	Y	N	N
7	0	N		0	PhD	Y	Y	Y
8	5	Y		2	MS	N	Y	Y
9	3	N		1	BS	N	Y	Y
10	15	Y		5	BS	N	N	Y
11	0	N		0	BS	N	N	N
12	1	N		1	PhD	Y	N	N
13	4	Y		1	BS	N	Y	Y
14	0	N		0	PhD	Y	N	Y

You can see that Excel actually imported this into a table, but if you were to look at the raw text you'd see that it's made up of comma-separated values.

The first line is the actual headings of each column, so what we have above are the number of years of prior experience, is the candidate currently employed or not, number of previous employers, the level of education, whether they went to a top-tier school, whether they had an internship while they were in school, and finally, the target that we're trying to predict on, whether or not they got a job offer in the end of the day. Now, we need to read that information into an RDD so we can do something with it.

Let's go back to our script:

```
rawData = sc.textFile("e:/sundog-consult/udemy/datascience/PastHires.csv")
header = rawData.first()
rawData = rawData.filter(lambda x:x != header)
```

The first thing we need to do is read that CSV data in, and we're going to throw away that first row, because that's our header information, remember. So, here's a little trick for doing that. We start off by importing every single line from that file into a raw data RDD, and I could call that anything I want, but we're calling it `sc.textFile`. SparkContext has a `textFile` function that will take a text file and create a new RDD, where each entry, each line of the RDD, consists of one line of input.

Make sure you change the path to that file to wherever you actually installed it, otherwise it won't work.

Now, I'm going to extract the first line, the first row from that RDD, by using the `first` function. So, now the header RDD will contain one entry that is just that row of column headers. And now, look what's going on in the above code, I'm using `filter` on my original data that contains all of the information in that CSV file, and I'm defining a `filter` function that will only let lines through if that line is not equal to the contents of that initial header row. What I've done here is, I've taken my raw CSV file and I've stripped out the first line by only allowing lines that do not equal that first line to survive, and I'm returning that back to the `rawData` RDD variable again. So, I'm taking `rawData`, filtering out that first line, and creating a new `rawData` that only contains the data itself. With me so far? It's not that complicated.

Now, we're going to use a map function. What we need to do next is start to make more structure out of this information. Right now, every row of my RDD is just a line of text, it is comma-delimited text, but it's still just a giant line of text, and I want to take that comma-separated value list and actually split it up into individual fields. At the end of the day, I want each RDD to be transformed from a line of text that has a bunch of information separated by commas into a Python list that has actual individual fields for each column of information that I have. So, that's what this lambda function does:

```
csvData = rawData.map(lambda x: x.split(","))
```

It calls the built-in Python function `split`, which will take a row of input, and split it on comma characters, and divide that into a list of every field delimited by commas.

The output of this map function, where I passed in a lambda function that just splits every line into fields based on commas, is a new RDD called `csvData`. And, at this point, `csvData` is an RDD that contains, on every row, a list where every element is a column from my source data. Now, we're getting close.

It turns out that in order to use a decision tree with MLlib, a couple of things need to be true. First of all, the input has to be in the form of LabeledPoint data types, and it all has to be numeric in nature. So, we need to transform all of our raw data into data that can actually be consumed by MLlib, and that's what the `createLabeledPoints` function that we skipped past earlier does. We'll get to that in just a second, first here's the call to it:

```
trainingData = csvData.map(createLabeledPoints)
```

We're going to call a map on `csvData`, and we are going to pass it the `createLabeledPoints` function, which will transform every input row into something even closer to what we want at the end of the day. So, let's look at what `createLabeledPoints` does:

```
def createLabeledPoints(fields):
    yearsExperience = int(fields[0])
    employed = binary(fields[1])
    previousEmployers = int(fields[2])
    educationLevel = mapEducation(fields[3])
    topTier = binary(fields[4])
    interned = binary(fields[5])
    hired = binary(fields[6])

    return LabeledPoint(hired, array([yearsExperience, employed,
        previousEmployers, educationLevel, topTier, interned]))
```

It takes in a list of fields, and just to remind you again what that looks like, let's pull up that `.csv` Excel file again:

	A	B	C	D	E	F	G
1	Years Exp	Employed	Previous e	Level of E	Top-tier s	Interned	Hired
2	10	Y	4	BS	N	N	Y
3	0	N	0	BS	Y	Y	Y
4	7	N	6	BS	N	N	N
5	2	Y	1	MS	Y	N	Y
6	20	N	2	PhD	Y	N	N
7	0	N	0	PhD	Y	Y	Y
8	5	Y	2	MS	N	Y	Y
9	3	N	1	BS	N	Y	Y
10	15	Y	5	BS	N	N	Y
11	0	N	0	BS	N	N	N
12	1	N	1	PhD	Y	N	N
13	4	Y	1	BS	N	Y	Y
14	0	N	0	PhD	Y	N	Y

So, at this point, every RDD entry has a field, it's a Python list, where the first element is the years of experience, second element is employed, so on and so forth. The problems here are that we want to convert those lists to LabeledPoints, and we want to convert everything to numerical data. So, all these yes and no answers need to be converted to ones and zeros. These levels of experience need to be converted from names of degrees to some numeric ordinal value. Maybe we'll assign the value zero to no education, one can mean BS, two can mean MS, and three can mean PhD, for example. Again, all these yes/no values need to be converted to zeros and ones, because at the end of the day, everything going into our decision tree needs to be numeric, and that's what `createLabeledPoints` does. Now, let's go back to the code and run through it:

```
def createLabeledPoints(fields):
    yearsExperience = int(fields[0])
    employed = binary(fields[1])
    previousEmployers = int(fields[2])
    educationLevel = mapEducation(fields[3])
    topTier = binary(fields[4])
    interned = binary(fields[5])
    hired = binary(fields[6])

    return LabeledPoint(hired, array([yearsExperience, employed,
        previousEmployers, educationLevel, topTier, interned]))
```

First, it takes in our list of `StringFields` ready to convert it into `LabeledPoints`, where the label is the target value-was this person hired or not? 0 or 1-followed by an array that consists of all the other fields that we care about. So, this is how you create a `LabeledPoint` that the `DecisionTree` MLlib class can consume. So, you see in the above code that we're converting years of experience from a string to an integer value, and for all the yes/no fields, we're calling this `binary` function, that I defined up at the top of the code, but we haven't discussed yet:

```
def binary(YN):
    if (YN == 'Y'):
        return 1
    else:
        return 0
```

All it does is convert the character yes to 1, otherwise it returns 0. So, Y will become 1, N will become 0. Similarly, I have a `mapEducation` function:

```
def mapEducation(degree):
    if (degree == 'BS'):
        return 1
    elif (degree =='MS'):
        return 2
    elif (degree == 'PhD'):
        return 3
    else:
        return 0
```

As we discussed earlier, this simply converts different types of degrees to an ordinal numeric value in exactly the same way as our yes/no fields.

As a reminder, this is the line of code that sent us running through those functions:

```
trainingData = csvData.map(createLabeledPoints)
```

At this point, after mapping our RDD using that `createLabeledPoints` function, we now have a `trainingData` RDD, and this is exactly what MLlib wants for constructing a decision tree.

Creating a test candidate and building our decision tree

Let's create a little test candidate we can use, so we can use our model to actually predict whether someone new would be hired or not. What we're going to do is create a test candidate that consists of an array of the same values for each field as we had in the CSV file:

```
testCandidates = [ array([10, 1, 3, 1, 0, 0])]
```

Let's quickly compare that code with the Excel document so you can see the array mapping:

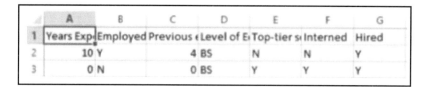

Again, we need to map these back to their original column representation, so that 10, 1, 3, 1, 0, 0 means 10 years of prior experience, currently employed, three previous employers, a BS degree, did not go to a top-tier school and did not do an internship. We could actually create an entire RDD full of candidates if we wanted to, but we'll just do one for now.

Next, we'll use parallelize to convert that list into an RDD:

```
testData = sc.parallelize(testCandidates)
```

Nothing new there. Alright, now for the magic let's move to the next code block:

```
model = DecisionTree.trainClassifier(trainingData, numClasses=2,
            categoricalFeaturesInfo={1:2, 3:4, 4:2, 5:2},
            impurity='gini', maxDepth=5, maxBins=32)
```

We are going to call `DecisionTree.trainClassifier`, and this is what will actually build our decision tree itself. We pass in our `trainingData`, which is just an RDD full of `LabeledPoint` arrays, `numClasses=2`, because we have, basically, a yes or no prediction that we're trying to make, will this person be hired or not? The next parameter is called `categoricalFeaturesInfo`, and this is a Python dictionary that maps fields to the number of categories in each field. So, if you have a continuous range available to a given field, like the number of years of experience, you wouldn't specify that at all in here, but for fields that are categorical in nature, such as what degree do they have, for example, that would say fieldID3, mapping to the degree attained, which has four different possibilities: no education, BS, MS, and PhD. For all of the yes/no fields, we're mapping those to 2 possible categories, yes/no or 0/1 is what we converted those to.

Chapter 9

Continuing to move through our `DecisionTree.trainClassifier` call, we are going to use the `'gini'` impurity metric as we measure the entropy. We have a `maxDepth` of 5, which is just an upper boundary on how far we're going to go, that can be larger if you wish. Finally, `maxBins` is just a way to trade off computational expense if you can, so it just needs to at least be the maximum number of categories you have in each feature. Remember, nothing really happens until we call an action, so we're going to actually use this model to make a prediction for our test candidate.

We use our `DecisionTree` model, which contains a decision tree that was trained on our test training data, and we tell that to make a prediction on our test data:

```
predictions = model.predict(testData)
print ('Hire prediction:')
results = predictions.collect()
for result in results:
    print (result)
```

We'll get back a list of predictions that we can then iterate through. So, `predict` returns a plain old Python object and is an action that I can `collect`. Let me rephrase that a little bit: `collect` will return a Python object on our predictions, and then we can iterate through every item in that list and print the result of the prediction.

We can also print out the decision tree itself by using `toDebugString`:

```
print('Learned classification tree model:')
print(model.toDebugString())
```

That will actually print out a little representation of the decision tree that it created internally, that you can follow through in your own head. So, that's kind of cool too.

Running the script

Alright, feel free to take some time, stare at this script a little bit more, digest what's going on, but, if you're ready, let's move on and actually run this beast. So, to do so, you can't just run it directly from Canopy. We're going to go to the Tools menu and open up a Canopy Command Prompt, and this just opens up a Windows command prompt with all the necessary environment variables in place for running Python scripts in Canopy. Make sure that the working directory is the directory that you installed all of the course materials into.

Apache Spark - Machine Learning on Big Data

All we need to do is call `spark-submit`, so this is a script that lets you run Spark scripts from Python, and then the name of the script, `SparkDecisionTree.py`. That's all I have to do.

```
spark-submit SparkDecisionTree.py
```

Hit Return, and off it will go. Again, if I were doing this on a cluster and I created my `SparkConf` accordingly, this would actually get distributed to the entire cluster, but, for now, we're just going to run it on my computer. When it's finished, you should see the below output:

```
Hire prediction:
1.0
Learned classification tree model:
DecisionTreeModel classifier of depth 4 with 9 nodes
  If (feature 1 in {0.0})
   If (feature 5 in {0.0})
    If (feature 0 <= 0.0)
     If (feature 3 in {1.0})
      Predict: 0.0
     Else (feature 3 not in {1.0})
      Predict: 1.0
    Else (feature 0 > 0.0)
     Predict: 0.0
   Else (feature 5 not in {0.0})
    Predict: 1.0
  Else (feature 1 not in {0.0})
   Predict: 1.0
```

So, in the above image, you can see in the test person that we put in above, we have a prediction that this person would be hired, and I've also printed out the decision tree itself, so it's kind of cool. Now, let's bring up that Excel document once more so we can compare it to the output:

	A	B	C	D	E	F	G
1	Years Exp	Employed	Previous employers	Level of Education	Top-tier school	Interned	Hired
2	10	Y		4	BS	N	N	Y
3	0	N	0	BS	Y	Y	Y
4	7	N	6	BS	N	N	N
5	2	Y	1	MS	Y	N	Y
6	20	N	2	PhD	Y	N	N
7	0	N	0	PhD	Y	Y	Y
8	5	Y	2	MS	N	Y	Y
9	3	N	1	BS	N	Y	Y
10	15	Y	5	BS	N	N	Y
11	0	N	0	BS	N	N	N
12	1	N	1	PhD	Y	N	N
13	4	Y	1	BS	N	Y	Y
14	0	N	0	PhD	Y	N	Y
15							

We can walk through this and see what it means. So, in our output decision tree we actually end up with a depth of four, with nine different nodes, and, again, if we remind ourselves what these different fields correlate to, the way to read this is: If (feature 1 in 0), so that means if the employed is No, then we drop down to feature 5. This list is zero-based, so feature 5 in our Excel document is internships. We can run through the tree like that: this person is not currently employed, did not do an internship, has no prior years of experience and has a Bachelor's degree, we would not hire this person. Then we get to the Else clauses. If that person had an advanced degree, we would hire them, just based on the data that we had that we trained it on. So, you can work out what these different feature IDs mean back to your original source data, remember, you always start counting at 0, and interpret that accordingly. Note that all the categorical features are expressed in Boolean in this list of possible categories that it saw, whereas continuous data is expressed numerically as less than or greater than relationships.

And there you have it, an actual decision tree built using Spark and MLlib that actually works and makes sense. Pretty awesome stuff.

K-Means Clustering in Spark

Alright, let's look at another example of using Spark in MLlib, and this time we're going to look at k-means clustering, and just like we did with decision trees, we're going to take the same example that we did using scikit-learn and we're going to do it in Spark instead, so it can actually scale up to a massive Dataset. So, again, I've made sure to close out of everything else, and I'm going to go into my book materials and open up the `SparkKMeans` Python script, and let's study what's going on in.

```
 7 K = 5
 8
 9 # Boilerplate Spark stuff:
10 conf = SparkConf().setMaster("local").setAppName("SparkKMeans")
11 sc = SparkContext(conf = conf)
12
13 #Create fake income/age clusters for N people in k clusters
14 def createClusteredData(N, k):
15     random.seed(10)
16     pointsPerCluster = float(N)/k
17     X = []
18     for i in range (k):
19         incomeCentroid = random.uniform(20000.0, 200000.0)
20         ageCentroid = random.uniform(20.0, 70.0)
21         for j in range(int(pointsPerCluster)):
22             X.append([random.normal(incomeCentroid, 10000.0), random.normal(ageCentroid, 2.0)])
23     X = array(X)
```

Alright, so again, we begin with some boilerplate stuff.

```
from pyspark.mllib.clustering import KMeans
from numpy import array, random
from math import sqrt
from pyspark import SparkConf, SparkContext
from sklearn.preprocessing import scale
```

We're going to import the `KMeans` package from the clustering `MLlib` package, we're going to import array and random from `numpy`, because, again, we're free to use whatever you want, this is a Python script at the end of the day, and `MLlib` often does require `numpy` arrays as input. We're going to import the `sqrt` function and the usual boilerplate stuff, we need `SparkConf` and `SparkContext`, pretty much every time from `pyspark`. We're also going to import the scale function from `scikit-learn`. Again, it's OK to use `scikit-learn` as long as you make sure its installed in every machine that you're going to be running this job on, and also don't assume that `scikit-learn` will magically scale itself up just because you're running it on Spark. But, since I'm only using it for the scaling function, it's OK. Alright, let's go ahead and set things up.

I'm going to create a global variable first:

```
K=5
```

I'm going to run k-means clustering in this example with a K of 5, meaning with five different clusters. I'm then going to go ahead and set up a local `SparkConf` just running on my own desktop:

```
conf = SparkConf().setMaster("local").setAppName("SparkKMeans")
sc = SparkContext(conf = conf)
```

Chapter 9

I'm going to set the name of my application to SparkKMeans and create a SparkContext object that I can then use to create RDDs that run on my local machine. We'll skip past the createClusteredData function for now, and go to the first line of code that gets run.

```
data = sc.parallelize(scale(createClusteredData(100, K)))
```

1. The first thing we're going to do is create an RDD by parallelizing in some fake data that I'm creating, and that's what the createClusteredData function does. Basically, I'm telling you to create 100 data points clustered around K centroids, and this is pretty much identical to the code that we looked at when we played with k-means clustering earlier in the book. If you want a refresher, go ahead and look back at that chapter. Basically, what we're going to do is create a bunch of random centroids around which we normally distribute some age and income data. So, what we're doing is trying to cluster people based on their age and income, and we are fabricating some data points to do that. That returns a numpy array of our fake data.

2. Once that result comes back from createClusteredData, I'm calling scale on it, and that will ensure that my ages and incomes are on comparable scales. Now, remember the section we studied saying you have to remember about data normalization? This is one of those examples where it is important, so we are normalizing that data with scale so that we get good results from k-means.

3. And finally, we parallelize the resulting list of arrays into an RDD using parallelize. Now our data RDD contains all of our fake data. All we have to do, and this is even easier than a decision tree, is call KMeans.train on our training data.

```
clusters = KMeans.train(data, K, maxIterations=10,
        initializationMode="random")
```

We pass in the number of clusters we want, our K value, a parameter that puts an upper boundary on how much processing it's going to do; we then tell it to use the default initialization mode of k-means where we just randomly pick our initial centroids for our clusters before we start iterating on them, and back comes the model that we can use. We're going to call that clusters.

Alright, now we can play with that cluster.

Apache Spark - Machine Learning on Big Data

Let's start by printing out the cluster assignments for each one of our points. So, we're going to take our original data and transform it using a lambda function:

```
resultRDD = data.map(lambda point: clusters.predict(point)).cache()
```

This function is just going to transform each point into the cluster number that is predicted from our model. Again, we're just taking our RDD of data points. We're calling `clusters.predict` to figure out which cluster our k-means model is assigning them to, and we're just going to put the results in our `resultRDD`. Now, one thing I want to point out here is this cache call, in the above code.

An important thing when you're doing Spark is that any time you're going to call more than one action on an RDD, it's important to cache it first, because when you call an action on an RDD, Spark goes off and figures out the DAG for it, and how to optimally get to that result.

It will go off and actually execute everything to get that result. So, if I call two different actions on the same RDD, it will actually end up evaluating that RDD twice, and if you want to avoid all of that extra work, you can cache your RDD in order to make sure that it does not recompute it more than once.

By doing that, we make sure these two subsequent operations do the right thing:

```
print ("Counts by value:")
counts = resultRDD.countByValue()
print (counts)

print ("Cluster assignments:")
results = resultRDD.collect()
print (results)
```

In order to get an actual result, what we're going to do is use `countByValue`, and what that will do is give us back an RDD that has how many points are in each cluster. Remember, `resultRDD` currently has mapped every individual point to the cluster it ended up with, so now we can use `countByValue` to just count up how many values we see for each given cluster ID. We can then easily print that list out. And we can actually look at the raw results of that RDD as well, by calling `collect` on it, and that will give me back every single points cluster assignment, and we can print out all of them.

Within set sum of squared errors (WSSSE)

Now, how do we measure how good our clusters are? Well, one metric for that is called the Within Set Sum of Squared Errors, wow, that sounds fancy! It's such a big term that we need an abbreviation for it, WSSSE. All it is, we look at the distance from each point to its centroid, the final centroid in each cluster, take the square of that error and sum it up for the entire Dataset. It's just a measure of how far apart each point is from its centroid. Obviously, if there's a lot of error in our model then they will tend to be far apart from the centroids that might apply, so for that we need a higher value of K, for example. We can go ahead and compute that value and print it out with the following code:

```
def error(point):
    center = clusters.centers[clusters.predict(point)]
    return sqrt(sum([x**2 for x in (point - center)]))

WSSSE = data.map(lambda point: error(point)).reduce(lambda x, y: x + y)
print("Within Set Sum of Squared Error = " + str(WSSSE))
```

First of all, we define this `error` function that computes the squared error for each point. It just takes the distance from the point to the centroid center of each cluster and sums it up. To do that, we're taking our source data, calling a lambda function on it that actually computes the error from each centroid center point, and then we can chain different operations together here.

First, we call `map` to compute the error for each point. Then to get a final total that represents the entire Dataset, we're calling `reduce` on that result. So, we're doing `data.map` to compute the error for each point, and then `reduce` to take all of those errors and add them all together. And that's what the little lambda function does. This is basically a fancy way of saying, "I want you to add up everything in this RDD into one final result." `reduce` will take the entire RDD, two things at a time, and combine them together using whatever function you provide. The function I'm providing it above is "take the two rows that I'm combining together and just add them up."

If we do that throughout every entry of the RDD, we end up with a final summed-up total. It might seem like a little bit of a convoluted way to just sum up a bunch of values, but by doing it this way we are able to make sure that we can actually distribute this operation if we need to. We could actually end up computing the sum of one piece of the data on one machine, and a sum of a different piece over on another machine, and then take those two sums and combine them together into a final result. This `reduce` function is saying, how do I take any two intermediate results from this operation, and combine them together?

Again, feel free to take a moment and stare at this a little bit longer if you want it to sink in. Nothing really fancy going on here, but there are a few important points:

- We introduced the use of a cache if you want to make sure that you don't do unnecessary recomputations on an RDD that you're going to use more than once.
- We introduced the use of the `reduce` function.
- We have a couple of interesting mapper functions as well here, so there's a lot to learn from in this example.

At the end of the day, it will just do k-means clustering, so let's go ahead and run it.

Running the code

Go to the Tools menu, Canopy Command Prompt, and type in:

```
spark-submit SparkKMeans.py
```

Hit Return, and off it will go. In this situation, you might have to wait a few moments for the output to appear in front of you, but you should see something like this:

```
(Canopy 64bit) E:\sundog-consult\Udemy\DataScience>spark-submit SparkKMeans.py
Counts by value:
defaultdict(<type 'int'>, {0: 21, 1: 20, 2: 20, 3: 20, 4: 19})
Cluster assignments:
[3, 3, 3, 3, 3, 3, 3, 3, 3, 3, 3, 3, 3, 3, 3, 3, 3, 3, 3, 3, 3, 1, 1, 1, 1, 1, 1, 1
, 1, 1, 1, 1, 1, 1, 1, 1, 1, 1, 1, 1, 4, 4, 4, 4, 4, 4, 4, 4, 4, 4, 4, 4, 4, 4, 4,
4, 4, 4, 4, 4, 0, 2, 2, 2, 2, 2, 2, 2, 2, 2, 2, 2, 2, 2, 2, 2, 2, 2, 2, 2, 2,
0, 0, 0, 0, 0, 0, 0, 0, 0, 0, 0, 0, 0, 0, 0, 0, 0, 0, 0]
Within Set Sum of Squared Error = 19.9732765798

(Canopy 64bit) E:\sundog-consult\Udemy\DataScience>
```

It worked, awesome! So remember, the output that we asked for was, first of all, a count of how many points ended up in each cluster. So, this is telling us that cluster 0 had 21 points in it, cluster 1 had 20 points in it, and so on and so forth. It ended up pretty evenly distributed, so that's a good sign.

Next, we printed out the cluster assignments for each individual point, and, if you remember, the original data that fabricated this data did it sequentially, so it's actually a good thing that you see all of the 3s together, and all the 1s together, and all the 4s together, it looks like it started to get a little bit confused with the 0s and 2s, but by and large, it seems to have done a pretty good job of uncovering the clusters that we created the data with originally.

And finally, we computed the WSSSE metric, it came out to 19.97 in this example. So, if you want to play around with this a little bit, I encourage you to do so. You can see what happens to that error metric as you increase or decrease the values of K, and think about why that may be. You can also experiment with what happens if you don't normalize all the data, does that actually affect your results in a meaningful way? Is that actually an important thing to do? And you can also experiment with the `maxIterations` parameter on the model itself and get a good feel of what that actually does to the final results, and how important it is. So, feel free to mess around with it and experiment away. That's k-means clustering done with MLlib and Spark in a scalable manner. Very cool stuff.

TF-IDF

So, our final example of MLlib is going to be using something called Term Frequency Inverse Document Frequency, or TF-IDF, which is the fundamental building block of many search algorithms. As usual, it sounds complicated, but it's not as bad as it sounds.

So, first, let's talk about the concepts of TF-IDF, and how we might go about using that to solve a search problem. And what we're actually going to do with TF-IDF is create a rudimentary search engine for Wikipedia using Apache Spark in MLlib. How awesome is that? Let's get started.

TF-IDF stands for Term Frequency and Inverse Document Frequency, and these are basically two metrics that are closely interrelated for doing search and figuring out the relevancy of a given word to a document, given a larger body of documents. So, for example, every article on Wikipedia might have a term frequency associated with it, every page on the Internet could have a term frequency associated with it for every word that appears in that document. Sounds fancy, but, as you'll see, it's a fairly simple concept.

- **All Term Frequency** means is how often a given word occurs in a given document. So, within one web page, within one Wikipedia article, within one whatever, how common is a given word within that document? You know, what is the ratio of that word's occurrence rate throughout all the words in that document? That's it. That's all term frequency is.
- **Document frequency**, is the same idea, but this time it is the frequency of that word across the entire corpus of documents. So, how often does this word occur throughout all of the documents that I have, all the web pages, all of the articles on Wikipedia, whatever. For example, common words like "a" or "the" would have a very high document frequency, and I would expect them to also have a very high term frequency, but that doesn't necessarily mean they're relevant to a given document.

You can kind of see where we're going with this. So, let's say we have a very high term frequency and a very low document frequency for a given word. The ratio of these two things can give me a measure of the relevance of that word to the document. So, if I see a word that occurs very often in a given document, but not very often in the overall space of documents, then I know that this word probably conveys some special meaning to this particular document. It might convey what this document is actually about.

So, that's TF-IDF. It just stands for Term Frequency x Inverse Document Frequency, which is just a fancy way of saying term frequency over document frequency, which is just a fancy way of saying how often does this word occur in this document compared to how often it occurs in the entire body of documents? It's that simple.

TF-IDF in practice

In practice, there are a few little nuances to how we use this. For example, we use the actual log of the inverse document frequency instead of the raw value, and that's because word frequencies in reality tend to be distributed exponentially. So, by taking the log, we end up with a slightly better weighting of words, given their overall popularity. There are some limitations to this approach, obviously, one is that we basically assume a document is nothing more than a bagful of words, we assume there are no relationships between the words themselves. And, obviously, that's not always the case, and actually parsing them out can be a good part of the work, because you have to deal with things like synonyms and various tenses of words, abbreviations, capitalizations, misspellings, and so on. This gets back to the idea of cleaning your data being a large part of your job as a data scientist, and it's especially true when you're dealing with natural language processing stuff. Fortunately, there are some libraries out there that can help you with this, but it is a real problem and it will affect the quality of your results.

Another implementation trick that we use with TF-IDF is, instead of storing actual string words with their term frequencies and inverse document frequency, to save space and make things more efficient, we actually map every word to a numerical value, a hash value we call it. The idea is that we have a function that can take any word, look at its letters, and assign that, in some fairly well-distributed manner, to a set of numbers in a range. That way, instead of using the word "represented", we might assign that a hash value of 10, and we can then refer to the word "represented" as "10" from now on. Now, if the space of your hash values isn't large enough, you could end up with different words being represented by the same number, which sounds worse than it is. But, you know, you want to make sure that you have a fairly large hash space so that is unlikely to happen. Those are called hash collisions. They can cause issues, but, in reality, there's only so many words that people commonly use in the English language. You can get away with 100,000 or so and be just fine.

Doing this at scale is the hard part. If you want to do this over all of Wikipedia, then you're going to have to run this on a cluster. But for the sake of argument, we are just going to run this on our own desktop for now, using a small sample of Wikipedia data.

Using TF-IDF

How do we turn that into an actual search problem? Once we have TF-IDF, we have this measure of each word's relevancy to each document. What do we do with it? Well, one thing you could do is compute TF-IDF for every word that we encounter in the entire body of documents that we have, and then, let's say we want to search for a given term, a given word. Let's say we want to search for "what Wikipedia article in my set of Wikipedia articles is most relevant to Gettysburg?" I could sort all the documents by their TF-IDF score for Gettysburg, and just take the top results, and those are my search results for Gettysburg. That's it. Just take your search word, compute TF-IDF, take the top results. That's it.

Obviously, in the real world there's a lot more to search than that. Google has armies of people working on this problem and it's way more complicated in practice, but this will actually give you a working search engine algorithm that produces reasonable results. Let's go ahead and dive in and see how it all works.

Searching wikipedia with Spark MLlib

We're going to build an actual working search algorithm for a piece of Wikipedia using Apache Spark in MLlib, and we're going to do it all in less than 50 lines of code. This might be the coolest thing we do in this entire book!

Go into your course materials and open up the `TF-IDF.py` script, and that should open up Canopy with the following code:

```python
from pyspark import SparkConf, SparkContext
from pyspark.mllib.feature import HashingTF
from pyspark.mllib.feature import IDF

# Boilerplate Spark stuff:
conf = SparkConf().setMaster("local").setAppName("SparkTFIDF")
sc = SparkContext(conf = conf)

# Load documents (one per line).
rawData = sc.textFile("e:/sundog-consult/Udemy/DataScience/subset-small.tsv")
fields = rawData.map(lambda x: x.split("\t"))
documents = fields.map(lambda x: x[3].split(" "))

# Store the document names for later:
documentNames = fields.map(lambda x: x[1])

# Now hash the words in each document to their term frequencies:
hashingTF = HashingTF(100000)  #100K hash buckets just to save some memory
tf = hashingTF.transform(documents)

# At this point we have an RDD of sparse vectors representing each document,
# where each value maps to the term frequency of each unique hash value.

# Let's compute the TF*IDF of each term in each document:
tf.cache()
idf = IDF(minDocFreq=2).fit(tf)
tfidf = idf.transform(tf)
```

Now, step back for a moment and let it sink in that we're actually creating a working search algorithm, along with a few examples of using it in less than 50 lines of code here, and it's scalable. I could run this on a cluster. It's kind of amazing. Let's step through the code.

Import statements

We're going to start by importing the `SparkConf` and `SparkContext` libraries that we need for any Spark script that we run in Python, and then we're going to import `HashingTF` and `IDF` using the following commands.

```
from pyspark import SparkConf, SparkContext
from pyspark.mllib.feature import HashingTF
from pyspark.mllib.feature import IDF
```

So, this is what computes the term frequencies (`TF`) and inverse document frequencies (`IDF`) within our documents.

Creating the initial RDD

We'll start off with our boilerplate Spark stuff that creates a local `SparkConfiguration` and a `SparkContext`, from which we can then create our initial RDD.

```
conf = SparkConf().setMaster("local").setAppName("SparkTFIDF")
sc = SparkContext(conf = conf)
```

Next, we're going to use our `SparkContext` to create an RDD from `subset-small.tsv`.

```
rawData = sc.textFile("e:/sundog-consult/Udemy/DataScience/subset-small.tsv")
```

This is a file containing tab-separated values, and it represents a small sample of Wikipedia articles. Again, you'll need to change your path as shown in the preceding code as necessary for wherever you installed the course materials for this book.

That gives me back an RDD where every document is in each line of the RDD. The `tsv` file contains one entire Wikipedia document on every line, and I know that each one of those documents is split up into tabular fields that have various bits of metadata about each article.

The next thing I'm going to do is split those up:

```
fields = rawData.map(lambda x: x.split("\t"))
```

I'm going to split up each document based on their tab delimiters into a Python list, and create a new `fields` RDD that, instead of raw input data, now contains Python lists of each field in that input data.

Apache Spark - Machine Learning on Big Data

Finally, I'm going to map that data, take in each list of fields, extract field number three `x[3]`, which I happen to know is the body of the article itself, the actual article text, and I'm in turn going to split that based on spaces:

```
documents = fields.map(lambda x: x[3].split(" "))
```

What `x[3]` does is extract the body of the text from each Wikipedia article, and split it up into a list of words. My new `documents` RDD has one entry for every document, and every entry in that RDD contains a list of words that appear in that document. Now, we actually know what to call these documents later on when we're evaluating the results.

I'm also going to create a new RDD that stores the document names:

```
documentNames = fields.map(lambda x: x[1])
```

All that does is take that same `fields` RDD and uses this `map` function to extract the document name, which I happen to know is in field number one.

So, I now have two RDDs, `documents`, which contains lists of words that appear in each document, and `documentNames`, which contains the name of each document. I also know that these are in the same order, so I can actually combine these together later on to look up the name for a given document.

Creating and transforming a HashingTF object

Now, the magic happens. The first thing we're going to do is create a `HashingTF` object, and we're going to pass in a parameter of 100,000. This means that I'm going to hash every word into one of 100,000 numerical values:

```
hashingTF = HashingTF(100000)
```

Instead of representing words internally as strings, which is very inefficient, it's going to try to, as evenly as possible, distribute each word to a unique hash value. I'm giving it up to 100,000 hash values to choose from. Basically, this is mapping words to numbers at the end of the day.

Next, I'm going to call `transform` on `hashingTF` with my actual RDD of documents:

```
tf = hashingTF.transform(documents)
```

That's going to take my list of words in every document and convert it to a list of hash values, a list of numbers that represent each word instead.

This is actually represented as a sparse vector at this point to save even more space. So, not only have we converted all of our words to numbers, but we've also stripped out any missing data. In the event that a word does not appear in a document where you're not storing the fact that word does not appear explicitly, it saves even more space.

Computing the TF-IDF score

To actually compute the TF-IDF score for each word in each document, we first cache this `tf` RDD.

```
tf.cache()
```

We do that because we're going to use it more than once. Next, we use `IDF(minDocFreq=2)`, meaning that we're going to ignore any word that doesn't appear at least twice:

```
idf = IDF(minDocFreq=2).fit(tf)
```

We call `fit` on `tf`, and then in the next line we call `transform` on `tf`:

```
tfidf = idf.transform(tf)
```

What we end up with here is an RDD of the TF-IDF score for each word in each document.

Using the Wikipedia search engine algorithm

Let's try and put the algorithm to use. Let's try to look up the best article for the word **Gettysburg**. If you're not familiar with US history, that's where Abraham Lincoln gave a famous speech. So, we can transform the word Gettysburg into its hash value using the following code:

```
gettysburgTF = hashingTF.transform(["Gettysburg"])
gettysburgHashValue = int(gettysburgTF.indices[0])
```

We will then extract the TF-IDF score for that hash value into a new RDD for each document:

```
gettysburgRelevance = tfidf.map(lambda x: x[gettysburgHashValue])
```

What this does is extract the TF-IDF score for Gettysburg, from the hash value it maps to for every document, and stores that in this `gettysburgRelevance` RDD.

We then combine that with the `documentNames` so we can see the results:

```
zippedResults = gettysburgRelevance.zip(documentNames)
```

Finally, we can print out the answer:

```
print ("Best document for Gettysburg is:")
print (zippedResults.max())
```

Running the algorithm

So, let's go run that and see what happens. As usual, to run the Spark script, we're not going to just hit the play icon. We have to go to **Tools>Canopy Command Prompt**. In the Command Prompt that opens up, we will type in `spark-submit TF-IDF.py`, and off it goes.

We are asking it to chunk through quite a bit of data, even though it's a small sample of Wikipedia it's still a fair chunk of information, so it might take a while. Let's see what comes back for the best document match for Gettysburg, what document has the highest TF-IDF score?

```
Best document for Gettysburg is:
(29.777067781559442, u'Abraham Lincoln')
```

It's Abraham Lincoln! Isn't that awesome? We just made an actual search engine that actually works, in just a few lines of code.

And there you have it, an actual working search algorithm for a little piece of Wikipedia using Spark in MLlib and TF-IDF. And the beauty is we can actually scale that up to all of Wikipedia if we wanted to, if we had a cluster large enough to run it.

Hopefully we got your interest up there in Spark, and you can see how it can be applied to solve what can be pretty complicated machine learning problems in a distributed manner. So, it's a very important tool, and I want to make sure you don't get through this book on data science without at least knowing the concepts of how Spark can be applied to big data problems. So, when you need to move beyond what one computer can do, remember, Spark is at your disposal.

Using the Spark 2.0 DataFrame API for MLlib

This chapter was originally produced for Spark 1, so let's talk about what's new in Spark 2, and what new capabilities exist in MLlib now.

So, the main thing with Spark 2 is that they moved more and more toward Dataframes and Datasets. Datasets and Dataframes are kind of used interchangeably sometimes. Technically a dataframe is a Dataset of row objects, they're kind of like RDDs, but the only difference is that, whereas an RDD just contains unstructured data, a Dataset has a defined schema to it.

A Dataset knows ahead of time exactly what columns of information exists in each row, and what types those are. Because it knows about the actual structure of that Dataset ahead of time, it can optimize things more efficiently. It also lets us think of the contents of this Dataset as a little, mini database, well, actually, a very big database if it's on a cluster. That means we can do things like issue SQL queries on it.

This creates a higher-level API with which we can query and analyze massive Datasets on a Spark cluster. It's pretty cool stuff. It's faster, it has more opportunities for optimization, and it has a higher-level API that's often easier to work with.

How Spark 2.0 MLlib works

Going forward in Spark 2.0, MLlib is pushing dataframes as its primary API. This is the way of the future, so let's take a look at how it works. I've gone ahead and opened up the `SparkLinearRegression.py` file in Canopy, as shown in the following figure, so let's walk through it a little bit:

```
1  from __future__ import print_function
2
3  from pyspark.ml.regression import LinearRegression
4
5  from pyspark.sql import SparkSession
6  from pyspark.ml.linalg import Vectors
7
8  if __name__ == "__main__":
9
10     # Create a SparkSession (Note, the config section is only for Windows!)
11     spark = SparkSession.builder.config("spark.sql.warehouse.dir", "file:///C:/temp").appName("LinearRegression").getOrCreate()
12
13     # Load up our data and convert it to the format MLLib expects.
14     inputLines = spark.sparkContext.textFile("regression.txt")
15     data = inputLines.map(lambda x: x.split(",")).map(lambda x: (float(x[0]), Vectors.dense(float(x[1]))))
16
17     # Convert this RDD to a DataFrame
18     colNames = ["label", "features"]
19     df = data.toDF(colNames)
20
21     # Note, there are lots of cases where you can avoid going from an RDD to a DataFrame.
22     # Perhaps you're importing data from a real database. Or you are using structured streaming
```

As you see, for one thing, we're using `ml` instead of `MLlib`, and that's because the new dataframe-based API is in there.

Implementing linear regression

In this example, what we're going to do is implement linear regression, and linear regression is just a way of fitting a line to a set of data. What we're going to do in this exercise is take a bunch of fabricated data that we have in two dimensions, and try to fit a line to it with a linear model.

We're going to separate our data into two sets, one for building the model and one for evaluating the model, and we'll compare how well this linear model does at actually predicting real values. First of all, in Spark 2, if you're going to be doing stuff with the `SparkSQL` interface and using Datasets, you've got to be using a `SparkSession` object instead of a `SparkContext`. To set one up, you do the following:

```
spark = SparkSession.builder.config("spark.sql.warehouse.dir",
"file:///C:/temp").appName("LinearRegression").getOrCreate()
```

> Note that the middle bit is only necessary on Windows and in Spark 2.0. It kind of works around a little bug that they have, to be honest. So, if you're on Windows, make sure you have a `C:/temp` folder. If you want to run this, go create that now if you need to. If you're not on Windows, you can delete that whole middle section to leave: `spark = SparkSession.builder.appName("LinearRegression").getOrCreate()`.

Okay, so you can say `spark`, give it an `appName` and `getOrCreate()`.

This is interesting, because once you've created a Spark session, if it terminates unexpectedly, you can actually recover from that the next time that you run it. So, if we have a checkpoint directory, it can actually restart where it left off using `getOrCreate`.

Now, we're going to use this `regression.txt` file that I have included with the course materials:

```
inputLines = spark.sparkContext.textFile("regression.txt")
```

That is just a text file that has comma-delimited values of two columns, and they're just two columns of, more or less randomly, linearly correlated data. It can represent whatever you want. Let's imagine that it represents heights and weights, for example. So, the first column might represent heights, the second column might represent weights.

In the lingo of machine learning, we talk about labels and features, where labels are usually the thing that you're trying to predict, and features are a set of known attributes of the data that you use to make a prediction from.

In this example, maybe heights are the labels and the features are the weights. Maybe we're trying to predict heights based on your weight. It can be anything, it doesn't matter. This is all normalized down to data between -1 and 1. There's no real meaning to the scale of the data anywhere, you can pretend it means anything you want, really.

To use this with MLlib, we need to transform our data into the format it expects:

```
data = inputLines.map(lambda x: x.split(",")).map(lambda x: (float(x[0]),
Vectors.dense(float(x[1]))))
```

The first thing we're going to do is split that data up with this map function that just splits each line into two distinct values in a list, and then we're going to map that to the format that MLlib expects. That's going to be a floating point label, and then a dense vector of the feature data.

In this case, we only have one bit of feature data, the weight, so we have a vector that just has one thing in it, but even if it's just one thing, the MLlib linear regression model requires a dense vector there. This is like a labeledPoint in the older API, but we have to do it the hard way here.

Next, we need to actually assign names to those columns. Here's the syntax for doing that:

```
colNames = ["label", "features"]
df = data.toDF(colNames)
```

We're going to tell MLlib that these two columns in the resulting RDD actually correspond to the label and the features, and then I can convert that RDD to a DataFrame object. At this point, I have an actual dataframe or, if you will, a Dataset that contains two columns, label and features, where the label is a floating point height, and the features column is a dense vector of floating point weights. That is the format required by MLlib, and MLlib can be pretty picky about this stuff, so it's important that you pay attention to these formats.

Apache Spark - Machine Learning on Big Data

Now, like I said, we're going to split our data in half.

```
trainTest = df.randomSplit([0.5, 0.5])
trainingDF = trainTest[0]
testDF = trainTest[1]
```

We're going to do a 50/50 split between training data and test data. This returns back two dataframes, one that I'm going to use to actually create my model, and one that I'm going to use to evaluate my model.

I will next create my actual linear regression model with a few standard parameters here that I've set.

```
lir = LinearRegression(maxIter=10, regParam=0.3, elasticNetParam=0.8)
```

We're going to call `lir = LinearRegression`, and then I will fit that model to the set of data that I held aside for training, the training data frame:

```
model = lir.fit(trainingDF)
```

That gives me back a model that I can use to make predictions from.

Let's go ahead and do that.

```
fullPredictions = model.transform(testDF).cache()
```

I will call `model.transform(testDF)`, and what that's going to do is predict the heights based on the weights in my testing Dataset. I actually have the known labels, the actual, correct heights, and this is going to add a new column to that dataframe called predictions, that has the predicted values based on that linear model.

I'm going to cache those results, and now I can just extract them and compare them together. So, let's pull out the prediction column, just using `select` like you would in SQL, and then I'm going to actually transform that dataframe and pull out the RDD from it, and use that to map it to just a plain old RDD full of floating point heights in this case:

```
predictions = fullPredictions.select("prediction").rdd.map(lambda x: x[0])
```

These are the predicted heights. Next, we're going to get the actual heights from the label column:

```
labels = fullPredictions.select("label").rdd.map(lambda x: x[0])
```

Finally, we can zip them back together and just print them out side by side and see how well it does:

```
predictionAndLabel = predictions.zip(labels).collect()

for prediction in predictionAndLabel:
    print(prediction)

spark.stop()
```

This is kind of a convoluted way of doing it; I did this to be more consistent with the previous example, but a simpler approach would be to just actually select prediction and label together into a single RDD that maps out those two columns together and then I don't have to zip them up, but either way it works. You'll also note that right at the end there we need to stop the Spark session.

So let's see if it works. Let's go up to Tools, Canopy Command Prompt, and we'll type in `spark-submit SparkLinearRegression.py` and let's see what happens.

There's a little bit more upfront time to actually run these APIs with Datasets, but once they get going, they're very fast. Alright, there you have it.

```
(0.8643234408131227, 1.2)
(0.9571202568137612, 1.31)
(0.9428438235828938, 1.34)
(1.0499170728143994, 1.36)
(0.9571202568137612, 1.38)
(1.057055289429833, 1.41)
(0.9142909571211588, 1.44)
(0.9285673903520263, 1.47)
(1.1212992389687366, 1.5)
(0.8928763072748577, 1.5)
(1.2569253546619772, 1.53)
(1.135575672199604, 1.53)
(1.1070228057378693, 1.53)
(1.2355107048156762, 1.54)
(1.1498521054304716, 1.55)
(1.164128538661339, 1.56)
(1.135575672199604, 1.59)
(1.1784049718922063, 1.61)
(1.392551470355218, 1.78)
(1.214096054969375, 1.8)
(1.264063571277411, 1.82)
(1.342583954047182, 1.86)
(1.4924865029712902, 2.09)
```

Here we have our actual and predicted values side by side, and you can see that they're not too bad. They tend to be more or less in the same ballpark. There you have it, a linear regression model in action using Spark 2.0, using the new dataframe-based API for MLlib. More and more, you'll be using these APIs going forward with MLlib in Spark, so make sure you opt for these when you can. Alright, that's MLlib in Spark, a way of actually distributing massive computing tasks across an entire cluster for doing machine learning on big Datasets. So, good skill to have. Let's move on.

Summary

In this chapter, we started with installing Spark, then moved to introducing Spark in depth while understanding how Spark works in combination with RDDs. We also walked through various ways of creating RDDs while exploring different operations. We then introduced MLlib, and stepped through some detailed examples of decision trees and K-Means Clustering in Spark. We then pulled off our masterstroke of creating a search engine in just a few lines of code using TF-IDF. Finally, we looked at the new features of Spark 2.0.

In the next chapter, we'll take a look at A/B testing and experimental design.

10
Testing and Experimental Design

In this chapter, we'll see the concept of A/B testing. We'll go through the t-test, the t-statistic, and the p-value, all useful tools for determining whether a result is actually real or a result of random variation. We'll dive into some real examples and get our hands dirty with some Python code and compute the t-statistics and p-values.

Following that, we'll look into how long you should run an experiment for before reaching a conclusion. Finally, we'll discuss the potential issues that can harm the results of your experiment and may cause you to reach the wrong conclusion.

We'll cover the following topics:

- A/B testing concepts
- T-test and p-value
- Measuring t-statistics and p-values using Python
- Determining how long to run an experiment
- A/B test gotchas

A/B testing concepts

If you work as a data scientist at a web company, you'll probably be asked to spend some time analyzing the results of A/B tests. These are basically controlled experiments on a website to measure the impact of a given change. So, let's talk about what A/B tests are and how they work.

A/B tests

If you're going to be a data scientist at a big tech web company, this is something you're going to definitely be involved in, because people need to run experiments to try different things on a website and measure the results of it, and that's actually not as straightforward as most people think it is.

What is an A/B test? Well, it's a controlled experiment that you usually run on a website, it can be applied to other contexts as well, but usually we're talking about a website, and we're going to test the performance of some change to that website, versus the way it was before.

You basically have a *control* set of people that see the old website, and a *test* group of people that see the change to the website, and the idea is to measure the difference in behavior between these two groups and use that data to actually decide whether this change was beneficial or not.

For example, I own a business that has a website, we license software to people, and right now I have a nice, friendly, orange button that people click on when they want to buy a license as shown on the left in the following figure. But what would happen if I changed the color of that button to blue, as shown on the right?

So in this example, if I want to find out whether blue would be better. How do I know?

I mean, intuitively, maybe that might capture people's attention more, or intuitively, maybe people are more used to seeing orange buy buttons and are more likely to click on that, I could spin that either way, right? So, my own internal biases or preconceptions don't really matter. What matters is how people react to this change on my actual website, and that's what an A/B test does.

A/B testing will split people up into people who see the orange button, and people who see the blue button, and I can then measure the behavior between these two groups and how they might differ, and make my decision on what color my buttons should be based on that data.

You can test all sorts of things with an A/B test. These include:

- **Design changes**: These can be changes in the color of a button, the placement of a button, or the layout of the page.
- **UI flow**: So, maybe you're actually changing the way that your purchase pipeline works and how people check out on your website, and you can actually measure the effect of that.
- **Algorithmic changes**: Let's consider the example of doing movie recommendations that we discussed in `Chapter 6`, *Recommender Systems*. Maybe I want to test one algorithm versus another. Instead of relying on error metrics and my ability to do a train test, what I really care about is driving purchases or rentals or whatever it is on this website.
    - The A/B test can let me directly measure the impact of this algorithm on the end result that I actually care about, and not just my ability to predict movies that other people have already seen.
    - And anything else you can dream up too, really, any change that impacts how users interact with your site is worth testing. Maybe it's even, making the website faster, or it could be anything.
- **Pricing changes**: This one gets a little bit controversial. You know, in theory, you can experiment with different price points using an A/B test and see if it actually increases volume to offset for the price difference or whatever, but use that one with caution.
    - If customers catch wind that other people are getting better prices than they are for no good reason, they're not going to be very happy with you. Keep in mind, doing pricing experiments can have a negative backlash and you don't want to be in that situation.

Measuring conversion for A/B testing

The first thing you need to figure out when you're designing an experiment on a website is what are you trying to optimize for? What is it that you really want to drive with this change? And this isn't always a very obvious thing. Maybe it's the amount that people spend, the amount of revenue. Well, we talked about the problems with variance in using amount spent, but if you have enough data, you can still, reach convergence on that metric a lot of times.

However, maybe that's not what you actually want to optimize for. Maybe you're actually selling some items at a loss intentionally just to capture market share. There's more complexity that goes into your pricing strategy than just top-line revenue.

Maybe what you really want to measure is profit, and that can be a very tricky thing to measure, because a lot of things cut into how much money a given product might make and those things might not always be obvious. And again, if you have loss leaders, this experiment will discount the effect that those are supposed to have. Maybe you just care about driving ad clicks on your website, or order quantities to reduce variance, maybe people are okay with that.

The bottom line is that you have to talk to the business owners of the area that's being tested and figure out what it is they're trying to optimize for. What are they being measured on? What is their success measured on? What are their key performance indicators or whatever the NBAs want to call it? And make sure that we're measuring the thing that it matters to them.

You can measure more than one thing at once too, you don't have to pick one, you can actually report on the effect of many different things:

- Revenue
- Profit
- Clicks
- Ad views

If these things are all moving in the right direction together, that's a very strong sign that this change had a positive impact in more ways than one. So, why limit yourself to one metric? Just make sure you know which one matters the most in what's going to be your criteria for success of this experiment ahead of time.

How to attribute conversions

Another thing to watch out for is attributing conversions to a change downstream. If the action you're trying to drive doesn't happen immediately upon the user experiencing the thing that you're testing, things get a little bit dodgy.

Let's say I change the color of a button on page A, the user then goes to page B and does something else, and ultimately buys something from page C.

Well, who gets credit for that purchase? Is it page A, or page B, or something in-between? Do I discount the credit for that conversion depending on how many clicks that person took to get to the conversion action? Do I just discard any conversion action that doesn't happen immediately after seeing that change? These are complicated things and it's very easy to produce misleading results by fudging how you account for these different distances between the conversion and the change that you're measuring.

Variance is your enemy

Another thing that you need to really internalize is that variance is your enemy when you're running an A/B test.

A very common mistake people make who don't know what they're doing with data science is that they will put up a test on a web page, blue button versus orange button, whatever it is, run it for a week, and take the mean amount spent from each of those groups. They then say "oh look! The people with the blue button on average spent a dollar more than the people with the orange button; blue is awesome, I love blue, I'm going to put blue all over the website now!"

But, in fact, all they might have been seeing was just a random variation in purchases. They didn't have a big enough sample because people don't tend to purchase a lot. You get a lot of views but you probably don't have a lot of purchases on your website in comparison, and it's probably a lot of variance in those purchase amounts because different products cost different amounts.

So, you could very easily end up making the wrong decision that ends up costing your company money in the long run, instead of earning your company money if you don't understand the effect of variance on these results. We'll talk about some principal ways of measuring and accounting for that later in the chapter.

You need to make sure that your business owners understand that this is an important effect that you need to quantify and understand before making business decisions following an A/B test or any experiment that you run on the web.

Now, sometimes you need to choose a conversion metric that has less variance. It could be that the numbers on your website just mean that you would have to run an experiment for years in order to get a significant result based on something like revenue or amount spent.

Sometimes if you're looking at more than one metric, such as order amount or order quantity, that has less variance associated with it, you might see a signal on order quantity before you see a signal on revenue, for example. At the end of the day, it ends up being a judgment call. If you see a significant lift in order quantities and maybe a not-so-significant lift in revenue, then you have to say "well, I think there might be something real and beneficial going on here."

However, the only thing that statistics and data size can tell you, are probabilities that an effect is real. It's up to you to decide whether or not it's real at the end of the day. So, let's talk about how to do this in more detail.

The key takeaway here is, just looking at the differences in means isn't enough. When you're trying to evaluate the results of an experiment, you need to take the variance into account as well.

T-test and p-value

How do you know if a change resulting from an A/B test is actually a real result of what you changed, or if it's just random variation? Well, there are a couple of statistical tools at our disposal called the t-test or t-statistic, and the p-value. Let's learn more about what those are and how they can help you determine whether an experiment is good or not.

The aim is to figure out if a result is real or not. Was this just a result of random variance that's inherent in the data itself, or are we seeing an actual, statistically significant change in behavior between our control group and our test group? T-tests and p-values are a way to compute that.

 Remember, *statistically significant* doesn't really have a specific meaning. At the end of the day it has to be a judgment call. You have to pick a probability value that you're going to accept of a result being real or not. But there's always going to be a chance that it's still a result of random variation, and you have to make sure your stakeholders understand that.

The t-statistic or t-test

Let's start with the **t-statistic**, also known as a t-test. It is basically a measure of the difference in behavior between these two sets, between your control and treatment group, expressed in units of standard error. It is based on standard error, which accounts for the variance inherent in the data itself, so by normalizing everything by that standard error, we get some measure of the change in behavior between these two groups that takes that variance into account.

The way to interpret a t-statistic is that a high t-value means there's probably a real difference between these two sets, whereas a low t-value means not so much difference. You have to decide what's a threshold that you're willing to accept? The sign of the t-statistic will tell you if it's a positive or negative change.

If you're comparing your control to your treatment group and you end up with a negative t-statistic, that implies that this is a bad change. You ultimate want the absolute value of that t-statistic to be large. How large a value of t-statistic is considered large? Well, that's debatable. We'll look at some examples shortly.

Now, this does assume that you have a normal distribution of behavior, and when we're talking about things like the amount people spend on a website, that's usually a decent assumption. There does tend to be a normal distribution of how much people spend.

However, there are more refined versions of t-statistics that you might want to look at for other specific situations. For example, there's something called **Fisher's exact test** for when you're talking about click through rates, the **E-test** when you're talking about transactions per user, like how many web pages do they see, and the **chi-squared** test, which is often relevant for when you're looking at order quantities. Sometimes you'll want to look at all of these statistics for a given experiment, and choose the one that actually fits what you're trying to do the best.

The p-value

Now, it's a lot easier to talk about p-values than t-statistics because you don't have to think about, how many standard deviations are we talking about? What does the actual value mean? The p-value is a little bit easier for people to understand, which makes it a better tool for you to communicate the results of an experiment to the stakeholders in your business.

The p-value is basically the probability that this experiment satisfies the null hypothesis, that is, the probability that there is no real difference between the control and the treatment's behavior. A low p-value means there's a low probability of it having no effect, kind of a double negative going on there, so it's a little bit counter intuitive, but at the end of the day you just have to understand that a low p-value means that there's a high probability that your change had a real effect.

What you want to see are a high t-statistic and a low p-value, and that will imply a significant result. Now, before you start your experiment, you need to decide what your threshold for success is going to be, and that means deciding the threshold with the people in charge of the business.

So, what p-value are you willing to accept as a measure of success? Is it 1 percent? Is it 5 percent? And again, this is basically the likelihood that there is no real effect, that it's just a result of random variance. It is just a judgment call at the end of the day. A lot of times people use 1 percent, sometimes they use 5 percent if they're feeling a little bit riskier, but there's always going to be that chance that your result was just spurious, random data that came in.

However, you can choose the probability that you're willing to accept as being likely enough that this is a real effect, that's worth rolling out into production.

When your experiment is over, and we'll talk about when you declare an experiment to be over later, you want to measure your p-value. If it's less than the threshold you decided upon, then you can reject the null hypothesis and you can say "well, there's a high likelihood that this change produced a real positive or negative result."

If it is a positive result then you can roll that change out to the entire site and it is no longer an experiment, it is part of your website that will hopefully make you more and more money as time goes on, and if it's a negative result, you want to get rid of it before it costs you any more money.

 Remember, there is a real cost to running an A/B test when your experiment has a negative result. So, you don't want to run it for too long because there's a chance you could be losing money.

This is why you want to monitor the results of an experiment on a daily basis, so if there are early indications that the change is making a horrible impact to the website, maybe there's a bug in it or something that's horrible, you can pull the plug on it prematurely if necessary, and limit the damage.

Let's go to an actual example and see how you might measure t-statistics and p-values using Python.

Measuring t-statistics and p-values using Python

Let's fabricate some experimental data and use the t-statistic and p-value to determine whether a given experimental result is a real effect or not. We're going to actually fabricate some fake experimental data and run t-statistics and p-values on them, and see how it works and how to compute it in Python.

Running A/B test on some experimental data

Let's imagine that we're running an A/B test on a website and we have randomly assigned our users into two groups, group A and group B. The A group is going to be our test subjects, our treatment group, and group B will be our control, basically the way the website used to be. We'll set this up with the following code:

```
import numpy as np
from scipy import stats

A = np.random.normal(25.0, 5.0, 10000)
B = np.random.normal(26.0, 5.0, 10000)

stats.ttest_ind(A, B)
```

In this code example, our treatment group (A) is going to have a randomly distributed purchase behavior where they spend, on average, $25 per transaction, with a standard deviation of five and ten thousand samples, whereas the old website used to have a mean of $26 per transaction with the same standard deviation and sample size. We're basically looking at an experiment that had a negative result. All you have to do to figure out the t-statistic and the p-value is use this handy `stats.ttest_ind` method from `scipy`. What you do is, you pass it in your treatment group and your control group, and out comes your t-statistic as shown in the output here:

```
Out[1]:  Ttest_indResult(statistic=-14.075196812141339, pvalue=8.8277957363196977e-45)
```

Testing and Experimental Design

In this case, we have a t-statistic of -14. The negative indicates that it is a negative change, this was a bad thing. And the p-value is very, very small. So, that implies that there is an extremely low probability that this change is just a result of random chance.

 Remember that in order to declare significance, we need to see a high t-value t-statistic, and a low p-value.

That's exactly what we're seeing here, we're seeing -14, which is a very high absolute value of the t-statistic, negative indicating that it's a bad thing, and an extremely low P-value, telling us that there's virtually no chance that this is just a result of random variation.

If you saw these results in the real world, you would pull the plug on this experiment as soon as you could.

When there's no real difference between the two groups

Just as a sanity check, let's go ahead and change things so that there's no real difference between these two groups. So, I'm going to change group B, the control group in this case, to be the same as the treatment, where the mean is 25, the standard deviation is unchanged, and the sample size is unchanged as shown here:

```
B = np.random.normal(25.0, 5.0, 10000)

stats.ttest_ind(A, B)
```

If we go ahead and run this, you can see our t-test ends up being below one now:

```
Out[2]: Ttest_indResult(statistic=0.088886198511817435, pvalue=0.92917324220169051)
```

Remember this is in terms of standard deviation. So this implies that there's probably not a real change there unless we have a much higher p-value as well, over 30 percent.

Now, these are still relatively high numbers. You can see that random variation can be kind of an insidious thing. This is why you need to decide ahead of time what would be an acceptable limit for p-value.

You know, you could look at this after the fact and say, "30 percent odds, you know, that's not so bad, we can live with that," but, no. I mean, in reality and practice you want to see p-values that are below 5 percent, ideally below 1 percent, and a value of 30 percent means it's actually not that strong of a result. So, don't justify it after the fact, go into your experiment in knowing what your threshold is.

Does the sample size make a difference?

Let's do some changes in the sample size. We're creating these sets under the same conditions. Let's see if we actually get a difference in behavior by increasing the sample size.

Sample size increased to six-digits

So, we're going to go from `10000` to `100000` samples as shown here:

```
A = np.random.normal(25.0, 5.0, 100000)
B = np.random.normal(25.0, 5.0, 100000)

stats.ttest_ind(A, B)
```

You can see in the following output that actually the p-value got a little bit lower and the t-test a little bit larger, but it's still not enough to declare a real difference. It's actually going in the direction you wouldn't expect it to go? Kind of interesting!

```
Out[6]: Ttest_indResult(statistic=0.20964627681745385, pvalue=0.83394397202032966)
```

But these are still high values. Again, it's just the effect of random variance, and it can have more of an effect than you realize. Especially on a website when you're talking about order amounts.

Sample size increased seven-digits

Let's actually increase the sample size to `1000000`, as shown here:

```
A = np.random.normal(25.0, 5.0, 1000000)
B = np.random.normal(25.0, 5.0, 1000000)

stats.ttest_ind(A, B)
```

Here is the result:

```
Out[9]: Ttest_indResult(statistic=-0.075342911693641518, pvalue=0.93994188742749496)
```

What does that do? Well, now, we're back under 1 for the t-statistic, and our value's around 35 percent.

We're seeing these kind of fluctuations a little bit in either direction as we increase the sample size. This means that going from 10,000 samples to 100,000 to 1,000,000 isn't going to change your result at the end of the day. And running experiments like this is a good way to get a good gut feel as to how long you might need to run an experiment for. How many samples does it actually take to get a significant result? And if you know something about the distribution of your data ahead of time, you can actually run these sorts of models.

A/A testing

If we were to compare the set to itself, this is called an A/A test as shown in the following code example:

```
stats.ttest_ind(A, A)
```

We can see in the following output, a t-statistic of 0 and a p-value of `1.0` because there is in fact no difference whatsoever between these sets.

```
Out[10]: Ttest_indResult(statistic=0.0, pvalue=1.0)
```

Now, if you were to run that using real website data where you were looking at the same exact people and you saw a different value, that indicates there's a problem in the system itself that runs your testing. At the end of the day, like I said, it's all a judgment call.

Go ahead and play with this, see what the effect of different standard deviations has on the initial datasets, or differences in means, and different sample sizes. I just want you to dive in, play around with these different datasets and actually run them, and see what the effect is on the t-statistic and the p-value. And hopefully that will give you a more gut feel of how to interpret these results.

Again, the important thing to understand is that you're looking for a large t-statistic and a small p-value. P-value is probably going to be what you want to communicate to the business. And remember, lower is better for p-value, you want to see that in the single digits, ideally below 1 percent before you declare victory.

We'll talk about A/B tests some more in the remainder of the chapter. SciPy makes it really easy to compute t-statistics and p-values for a given set of data, so you can very easily compare the behavior between your control and treatment groups, and measure what the probability is of that effect being real or just a result of random variation. Make sure you are focusing on those metrics and you are measuring the conversion metric that you care about when you're doing those comparisons.

Determining how long to run an experiment for

How long do you run an experiment for? How long does it take to actually get a result? At what point do you give up? Let's talk about that in more detail.

If someone in your company has developed a new experiment, a new change that they want to test, then they have a vested interest in seeing that succeed. They put a lot of work and time into it, and they want it to be successful. Maybe you've gone weeks with the testing and you still haven't reached a significant outcome on this experiment, positive or negative. You know that they're going to want to keep running it pretty much indefinitely in the hope that it will eventually show a positive result. It's up to you to draw the line on how long you're willing to run this experiment for.

How do I know when I'm done running an A/B test? I mean, it's not always straightforward to predict how long it will take before you can achieve a significant result, but obviously if you have achieved a significant result, if your p-value has gone below 1 percent or 5 percent or whatever threshold you've chosen, and you're done.

Testing and Experimental Design

At that point you can pull the plug on the experiment and either roll out the change more widely or remove it because it was actually having a negative effect. You can always tell people to go back and try again, use what they learned from the experiment to maybe try it again with some changes and soften the blow a little bit.

The other thing that might happen is it's just not converging at all. If you're not seeing any trends over time in the p-value, it's probably a good sign that you're not going to see this converge anytime soon. It's just not going to have enough of an impact on behavior to even be measurable, no matter how long you run it.

In those situations, what you want to do every day is plot on a graph for a given experiment the p-value, the t-statistic, whatever you're using to measure the success of this experiment, and if you're seeing something that looks promising, you will see that p-value start to come down over time. So, the more data it gets, the more significant your results should be getting.

Now, if you instead see a flat line or a line that's all over the place, that kind of tells you that that p-value's not going anywhere, and it doesn't matter how long you run this experiment, it's just not going to happen. You need to agree up front that in the case where you're not seeing any trends in p-values, what's the longest you're willing to run this experiment for? Is it two weeks? Is it a month?

Another thing to keep in mind is that having more than one experiment running on the site at once can conflate your results.

Time spent on experiments is a valuable commodity, you can't make more time in the world. You can only really run as many experiments as you have time to run them in a given year. So, if you spend too much time running one experiment that really has no chance of converging on a result, that's an opportunity you've missed to run another potentially more valuable experiment during that time that you are wasting on this other one.

It's important to draw the line on experiment links, because time is a very precious commodity when you're running A/B tests on a website, at least as long as you have more ideas than you have time, which hopefully is the case. Make sure you go in with agreed upper bounds on how long you're going to spend testing a given experiment, and if you're not seeing trends in the p-value that look encouraging, it's time to pull the plug at that point.

A/B test gotchas

An important point I want to make is that the results of an A/B test, even when you measure them in a principled manner using p-values, is not gospel. There are many effects that can actually skew the results of your experiment and cause you to make the wrong decision. Let's go through a few of these and let you know how to watch out for them. Let's talk about some gotchas with A/B tests.

It sounds really official to say there's a p-value of 1 percent, meaning there's only a 1 percent chance that a given experiment was due to spurious results or random variation, but it's still not the be-all and end-all of measuring success for an experiment. There are many things that can skew or conflate your results that you need to be aware of. So, even if you see a p-value that looks very encouraging, your experiment could still be lying to you, and you need to understand the things that can make that happen so you don't make the wrong decisions.

Remember, correlation does not imply causation.

Even with a well-designed experiment, all you can say is there is some probability that this effect was caused by this change you made.

At the end of the day, there's always going to be a chance that there was no real effect, or you might even be measuring the wrong effect. It could still be random chance, there could be something else going on, it's your duty to make sure the business owners understand that these experimental results need to be interpreted, they need to be one piece of their decision.

They can't be the be-all and end-all that they base their decision on because there is room for error in the results and there are things that can skew those results. And if there's some larger business objective to this change, beyond just driving short-term revenue, that needs to be taken into account as well.

Novelty effects

One problem is novelty effects. One major Achilles heel of an A/B test is the short time frame over which they tend to be run, and this causes a couple of problems. First of all, there might be longer-term effects to the change, and you're not going to measure those, but also, there is a certain effect to just something being different on the website.

For instance, maybe your customers are used to seeing the orange buttons on the website all the time, and if a blue button comes up and it catches their attention just because it's different. However, as new customers come in who have never seen your website before, they don't notice that as being different, and over time even your old customers get used to the new blue button. It could very well be that if you were to make this same test a year later, there would be no difference. Or maybe they'd be the other way around.

I could very easily see a situation where you test orange button versus blue button, and in the first two weeks the blue button wins. People buy more because they are more attracted to it, because it's different. But a year goes by, I could probably run another web lab that puts that blue button against an orange button and the orange button would win, again, simply because the orange button is different, and it's new and catches people's attention just for that reason alone.

For that reason, if you do have a change that is somewhat controversial, it's a good idea to rerun that experiment later on and see if you can actually replicate its results. That's really the only way I know of to account for novelty effects; actually measure it again when it's no longer novel, when it's no longer just a change that might capture people's attention simply because it's different.

And this, I really can't understate the importance of understanding this. This can really skew a lot of results, it biases you to attributing positive changes to things that don't really deserve it. Being different in and of itself is not a virtue; at least not in this context.

Seasonal effects

If you're running an experiment over Christmas, people don't tend to behave the same during Christmas as they do the rest of the year. They definitely spend their money differently during that season, they're spending more time with their families at home, and they might be a little bit, kind of checked out of work, so people have a different frame of mind.

It might even be involved with the weather, during the summer people behave differently because it's hot out they're feeling kind of lazy, they're on vacation more often. Maybe if you happen to do your experiment during the time of a terrible storm in a highly populated area that could skew your results as well.

Again, just be cognizant of potential seasonal effects, holidays are a big one to be aware of, and always take your experience with a grain of salt if they're run during a period of time that's known to have seasonality.

You can determine this quantitatively by actually looking at the metric you're trying to measure as a success metric, be it, whatever you're calling your conversion metric, and look at its behavior over the same time period last year. Are there seasonal fluctuations that you see every year? And if so, you want to try to avoid running your experiment during one of those peaks or valleys.

Selection bias

Another potential issue that can skew your results is selection bias. It's very important that customers are randomly assigned to either your control or your treatment groups, your A or B group.

However, there are subtle ways in which that random assignment might not be random after all. For example, let's say that you're hashing your customer IDs to place them into one bucket or the other. Maybe there's some subtle bias between how that hash function affects people with lower customer IDs versus higher customer IDs. This might have the effect of putting all of your longtime, more loyal customers into the control group, and your newer customers who don't know you that well into your treatment group.

What you end up measuring then is just a difference in behavior between old customers and new customers as a result. It's very important to audit your systems to make sure there is no selection bias in the actual assignment of people to the control or treatment group.

You also need to make sure that assignment is sticky. If you're measuring the effect of a change over an entire session, you want to measure if they saw a change on page A but, over on page C they actually did a conversion, you have to make sure they're not switching groups in between those clicks. So, you need to make sure that within a given session, people remain in the same group, and how to define a session can become kind of nebulous as well.

Now, these are all issues that using an established off-the-shelf framework like Google Experiments or Optimizely or one of those guys can help with so that you're not reinventing the wheel on all these problems. If your company does have a homegrown, in-house solution because they're not comfortable with sharing that data with outside companies, then it's worth auditing whether there is selection bias or not.

Auditing selection bias issues

One way for auditing selection bias issues is running what's called an A/A test, like we saw earlier. So, if you actually run an experiment where there is no difference between the treatment and control, you shouldn't see a difference in the end result. There should not be any sort of change in behavior when you're comparing those two things.

An A/A test can be a good way of testing your A/B framework itself and making sure there's no inherent bias or other problems, for example, session leakage and whatnot, that you need to address.

Data pollution

Another big problem is data pollution. We talked at length about the importance of cleaning your input data, and it's especially important in the context of an A/B test. What would happen if you have a robot, a malicious crawler that's crawling through your website all the time, doing an unnatural amount of transactions? What if that robot ends up getting either assigned to the treatment or the control?

That one robot could skew the results of your experiment. It's very important to study the input going into your experiment and look for outliers, then analyze what those outliers are, and whether they should they be excluded. Are you actually letting some robots leak into your measurements and are they skewing the results of your experiment? This is a very, very common problem, and something you need to be cognizant of.

There are malicious robots out there, there are people trying to hack into your website, there are benign scrapers just trying to crawl your website for search engines or whatnot. There are all sorts of weird behaviors going on with a website, and you need to filter out those and get at the people who are really your customers and not these automated scripts. That can actually be a very challenging problem. Yet another reason to use off-the-shelf frameworks like Google Analytics, if you can.

Attribution errors

We talked briefly about attribution errors earlier. This is if you are actually using downstream behavior from a change, and that gets into a gray area.

You need to understand how you're actually counting those conversions as a function of distance from the thing that you changed and agree with your business stakeholders upfront as to how you're going to measure those effects. You also need to be aware of if you're running multiple experiments at once; will they conflict with one another? Is there a page flow where someone might actually encounter two different experiments within the same session?

If so, that's going to be a problem and you have to apply your judgment as to whether these changes actually could interfere with each other in some meaningful way and affect the customers' behavior in some meaningful way. Again, you need to take these results with a grain of salt. There are a lot of things that can skew results and you need to be aware of them. Just be aware of them and make sure your business owners are also aware of the limitations of A/B tests and all will be okay.

Also, if you're not in a position where you can actually devote a very long amount of time to an experiment, you need to take those results with a grain of salt and ideally retest them later on during a different time period.

Summary

In this chapter, we talked about what A/B tests are and what are the challenges surrounding them. We went into some examples of how you actually measure the effects of variance using the t-statistic and p-value metrics, and we got into coding and measuring t-tests using Python. We then went on to discuss the short-term nature of an A/B test and its limitations, such as novelty effects or seasonal effects.

That also wraps up our time in this book. Congratulations for making it this far, that's a serious achievement and you should be proud of yourself. We've covered a lot of material here and I hope that you at least understand the concepts and have a little bit of hands-on experience with most of the techniques that are used in data science today. It's a very broad field, so we've touched on a little bit of everything there. So, you know, congratulations again.

If you want to further your career in this field, what I'd really encourage you to do is talk to your boss. If you work at a company that has access to some interesting datasets of its own, see if you can play around with them. Obviously, you want to talk to your boss first before you use any data owned by your company, because there's probably going to be some privacy restrictions surrounding it. You want to make sure that you're not violating the privacy of your company's customers, and that might mean that you might only be able to use that data or look at it within a controlled environment at your workplace. So, be careful when you're doing that.

If you can get permission to actually stay late at work a few days a week and, you know, mess around with some of these datasets and see what you can do with it, not only does show that you have the initiative to make yourself a better employee, you might actually discover something that might be valuable to your company, and that could just make you look even better, and actually lead to an internal transfer perhaps, into a field more directly related to where you want to take your career.

So, if you want some career advice from me, a common question I get is, "hey, I'm an engineer, I want to get more into data science, how do I do that?" The best way to do it is just do it, you know, actually do some side projects and show that you can do it and demonstrate some meaningful results from it. Show that to your boss and see where it leads you. Good luck.

Index

A

A/B test challenges
  about 393
  attribution errors 397
  data pollution 396
  novelty effects 394
  seasonal effects 394
  selection bias 395
  selection bias issues, auditing 396

A/B testing
  A/B tests 380
  about 379
  concepts 379
  conversion, measuring for 382
  conversions, attributing 383
  no difference between two groups 388
  performing 381
  variance, as enemy 383

Apache Spark
  components 336
  faster 335
  features 333
  high-level overview 333
  installing 310, 319, 320, 321, 322, 324, 326, 328, 331, 332
  installing, in other operating systems 311
  installing, on Windows 310
  Java Development Kit, installing 312, 313, 314, 316, 318
  k-means clustering 359
  Python, versus Scala 337
  relatively young 336
  Resilient Distributed Datasets (RDD) 338
  scalable 334
  user-friendly 336

axes
  adjusting 101
  labeling 107

B

bagging 199
bar charts
  generating 111
Bayes Optimal Classifier 200
Bayes' theorem 132
Bayesian methods 170, 171
Bayesian Model Combination 200
Bayesian Parameter Averaging 200
bias 276
bias-variance trade-off 276
binomial probability mass function 82
Boolean expression
  if statement 38
  if-else loop 38
boosting 199
bootstrap aggregating 199
box-and-whisker plots
  generating 113
bucket of models 199

C

categorical data 52
chi-squared test 385
conditional probability
  about 124
  assignment 129
  exercises, in Python 125, 126, 127, 128, 129
  my assignment solution 130
correlation
  about 118
  activity 124
  computing, NumPy way 122, 123
  computing, the hard way 118, 120, 121, 122

CountVectorizer 172
covariance and correlation
  about 116
  computing, in Python 118
  defining 116
covariance
  measuring 117

D

data cleaning 284
data distributions, types
  about 77
  exponential probability distribution 80
  Gaussian distribution 78
  Poisson probability mass function 83
  uniform distribution 77
data types, MLlib
  about 345
  LabeledPoint data type 346
  Rating data type 346
  vector data type 346
data types
  about 50
  categorical data 52
  numerical data 51
  ordinal data 53
data warehousing
  ETL, versus ELT 266
  overview 264
data, issues
  erroneous data 286
  formatting 286
  inconsistent data 286
  irrelevant data 286
  malicious data 285
  missing data 285
  outliers 285
decision trees code
  data, cleaning 351, 353, 355
  data, importing 351, 353, 355
  exploring 348
  script, running 357, 359
  SparkContext, creating 349
  test candidate, creating 356
decision trees, with MLlib

  about 347
  building 356
  decision trees code, exploring 348
decision trees
  concepts 186, 187
  example 188
  hiring decisions, predicting with Python 191,
    192, 193, 194, 196
  working with 190
dimensionality reduction 256
dynamic programming 272

E

E-test 385
ensemble learning
  about 198
  bagging 199
  boosting 199
  bootstrap aggregating 199
  bucket of models 199
  stacking 200
Enthought Canopy
  installation, testing 12, 14
  installing 8, 9, 11, 12
  IPNYB files issues, handling 15
entropy
  measuring 184, 185, 186
experiment
  length, determining 391
exploration problem
  about 270
  better way 271
  simple approach 270
exponential probability distribution
  about 80
  binomial probability mass function 82

F

Fisher's exact test 385
functions, in Python
  about 35
  Boolean expression syntax 37
  lambda functions 36

G

Gaussian distribution 78
Gettysburg 371
graphs
  saving, as images 100
GraphX 337
grid
  adding 103

H

histograms
  generating 113

I

ID3 (Iterative Dichotomiser 3) 190
interquartile range (IQR) 87
ipynb files 41
IPython Notebook
  about 15
  using 17, 18, 19
item-based collaborative filtering
  about 215
  advantages 216
  using 220
  working 217, 219

J

Jupyter Notebook 15

K

k-fold cross-validation
  about 164, 279
  example, scikit-learn used 280
  used, to avoid overfitting 279
k-means clustering, Spark
  about 359, 362
  code, running 364
  within set sum of squared errors (WSSSE) 363
k-means clustering
  about 177, 178
  example 181, 182, 183
  limitations 180
K-Nearest neighbors (KNN)
  about 246

  example 255
  used, for predicting movie rating 248, 252, 255

L

LabeledPoint data type 346
latent variable 161
line types and colors
  changing 103
linear regression
  about 135
  activity 143
  co-efficient of determination 139
  computing, Python used 140, 143
  gradient descent technique 138
  ordinary least squares technique 137
  r-squared 139
lists
  adding, to list 28
  append function 29
  complex data structures 29
  experimenting with 26
  negative syntax 28
  post colon 27
  pre colon 27
  reverse sort 30
  single element, dereferencing 29
  sort function 30
looping
  about 38, 39
  while loop 39

M

machine learning
  about 160
  supervised learning 162
  unsupervised learning 160, 161
Markov decision process 271
matplotlib
  axes, adjusting 101
  axes, labeling 107
  bar charts, generating 111
  box-and-whisker plots, generating 113
  crash course 98
  exercises 115
  fun example 108

graphs, saving as images 100
grid, adding 102
histograms, generating 113
legend, adding 107
line types and colors, changing 103, 106
multiple plots, generating on one graph 99
pie charts, generating 110
scatter plots, generating 112
mean
  about 55
  calculating, NumPy package used 58
  data, visualizing with matplotlib 59
  using, in Python 58
median
  about 54, 55
  calculating, NumPy package used 60
  effect of outliers, analyzing 61
  factor of outliers 56
  using, in Python 58
MLlib
  about 337, 344
  capabilities 345
  data types 345
mode
  about 54, 57
  calculating, SciPy package used 62, 64
  exercises 65
  using, in Python 58
modules
  data structures 26
  dictionaries 34
  importing 25
  iterating, through entries 34
  lists, experimenting with 26
  tuples 30
moments
  about 84, 90, 92
  computing, in Python 93
movie rating
  prediction, K-Nearest neighbors (KNN) used 248, 252, 255
movie recommendations
  entries with drop command, removing 241
  example 238
  groupby command, used for combining rows 240

  making, to people 233, 237
movie similarities
  code 222, 224
  corrwith function 225, 228
  finding 220
  results, improving 228, 231
multi-level models 155
multiple plots
  generating, on one graph 99
multivariate regression
  activity 154
  car prices, predicting 149
  Python used 151, 154

N

Naïve Bayes
  spam classifier, implementing with 172, 174, 175, 176
normal distribution 78
normalisation 284
numerical data
  about 51
  continuous data 51
  discrete data 51
  normalizing 301

O

ordinal data 53
ordinary least squares technique 137
outliers
  activity 307
  dealing with 304
  detecting 303

P

p-value 385, 386
percentiles
  about 84
  computing, in Python 87, 89
  quartiles 86
pie charts
  generating 110
Poisson probability mass function 83
polynomial regression
  about 144

activity 148
  implementing, NumPy used 145, 148
  r-squared error, computing 148
population variance
  versus sample variance 70
post requests
  filtering 293
Power Law 80
predictive models 135
principal component analysis
  about 256, 257, 259
  example, applying to Iris dataset 259, 264
probability density function 74, 75
probability mass function 74, 76
Python Basics
  activity, exploring 40
  functions 35
  looping 38
  part 1 20, 21
  part 2 35
Python code 22, 24
Python Markov Decision Process Toolbox
  reference 273
Python scripts
  running 41
  running, from Canopy IDE 44, 45
  running, in command prompt 43
  running, IPython/Jupyter Notebook used 42

R

r-squared
  computing 139
  computing, Python used 140, 143
  interpreting 139
random forests
  about 190, 191
  using 197
Rating data type 346
RDD operations
  about 341
  actions 343
  transformations 342
recommendations results
  improving 242
recommender systems

  about 210, 212
  user-based collaborative filtering 212
  user-based collaborative filtering, limitations 213
regular expression
  applying, on web log data 288
reinforcement learning
  about 268
  exploration problem 270
  fancy words 271
  Q-learning 269
request field
  filtering 291
Resilient Distributed Datasets (RDD)
  about 339
  creating 340
  creating, Python list used 340
  loading, from text file 340
  SparkContext object 339
  ways of creating 341

S

sample size
  A/A testing 390
  changing 389
  increasing, to seven-digits 390
  increasing, to six-digits 389
scatter plot
  about 112
  generating 112
scikit-learn
  using, for k-fold cross-validation example 280
shilling attack 215
spam classifier
  implementing, with Naïve Bayes 172, 174, 175, 176
Spark 2.0 DataFrame API
  using, for MLlib 373
Spark 2.0 MLlib
  linear regression, implementing 374, 375, 378
  working 373
Spark SQL 337
Spark Streaming 337
SparkContext object 339
stacking 200
standard deviation

about 65, 68
analyzing, on histogram 72
computing, Python used 73
example 74
outliers, identifying with 69
supervised learning
  about 162
  evaluating 162, 163
support vector machines (SVM)
  example 203, 206, 207
  overview 201

T

t-statistic 385
t-statistics and p-values
  A/B test, running on experimental data 387, 388
  measuring, Python used 387
t-test 384, 385
TF-IDF
  about 365
  All Term Frequency 365
  document frequency 365
  in practice 366
  using 367
TGZ (Tar in GZip) file 320
train/test
  about 145, 160
  used, for preventing overfitting of polynomial regression 164, 165, 166, 167, 168
transformations
  about 342
  Cartesian() 342
  distinct() 342
  filter() 342
  flatmap() 342
  intersection() 342
  map() 342
  map(), using 343
  sample() 342
  subtract() 342
  union() 342
tuples
  about 30

dictionaries 33
element, dereferencing 31
list of tuples 31, 32

U

uniform distribution 78
unsupervised learning 160, 161
user agents
  spiders/robots, activity filtering 297
  verifying 295
user-based collaborative filtering 212

V

variance
  about 65, 66, 276
  analyzing, on histogram 72
  computing, Python used 73
  example 74
  measuring 67
vector data type 346

W

web log data
  activity 301
  cleaning 287
  post requests, filtering 293
  regular expression, applying 288
  request field, filtering 291
  user agents, verifying 295
  website-specific filters, applying 299
website-specific filters
  applying 299
Wikipedia search engine algorithm
  running 372
  using 371
Wikipedia, searching with Spark MLlib
  about 367
  HashingTF object, creating 370
  HashingTF object, transforming 370
  initial RDD, creating 369
  statements, importing 369
  TF-IDF score, computing 371
  Wikipedia search engine algorithm, using 371